Workshop on Biomedical Natural Language Processing 2011

(BiomedNLP 2011)

Held with the 8th International Conference on Recent Advances in Natural Language Processing 2011 (RANLP 2011)

Hissar, Bulgaria
15 September 2011

ISBN: 978-1-62276-462-4

Printed from e-media with permission by:

Curran Associates, Inc.
57 Morehouse Lane
Red Hook, NY 12571

Some format issues inherent in the e-media version may also appear in this print version.

Printed by Curran Associates, Inc. (2012)

For permission requests, please contact the Association for Computational Linguistics
at the address below.

Association for Computational Linguistics
209 N. Eighth Street
Stroudsburg, Pennsylvania 18360

Phone: 1-570-476-8006
Fax: 1-570-476-0860

acl@aclweb.org

Additional copies of this publication are available from:

Curran Associates, Inc.
57 Morehouse Lane
Red Hook, NY 12571 USA
Phone: 845-758-0400
Fax: 845-758-2634
Email: curran@proceedings.com
Web: www.proceedings.com

Table of Contents

Assignment of Ontology-based Broad Semantic Classes to Biomedical Text (Invited)
Kevin Bretonnel Cohen 1

Exploiting Structured Data, Negation Detection and SNOMED CT Terms in a Random Indexing Approach to Clinical Coding
Aron Henriksson and Martin Hassel 3

Automatic Matching of ICD-10 codes to Diagnoses in Discharge Letters
Svetla Boytcheva 11

Evaluation Measures for Detection of Personal Health Information
Marina Sokolova 19

Building a Named Entity Recognizer in Three Days: Application to Disease Name Recognition in Bulgarian Epicrises
Georgi Georgiev, Valentin Zhivkov, Borislav Popov and Preslav Nakov 27

Reducing Complexity in Parsing Scientific Medical Data, a Diabetes Case Study
Dimitrios Kokkinakis 35

Architecture and Systems for Monitoring Hospital Acquired Infections inside Hospital Information Workflows
Denys Proux, Caroline Hagège, Quentin Gicquel, Suzanne Pereira, Stefan Darmoni, Frédérique Segond and Marie-Hélène Metzger 43

Towards Temporal Segmentation of Patient History in Discharge Letters
Galia Angelova and Svetla Boytcheva 49

Preface

Biomedical natural language processing deals with the application of text mining techniques to clinical documents and to scientific publications in the areas of biology and medicine. Recent years have seen dramatic changes in the types and amount of data available to researchers in this field. Where most research on publications in the past has dealt with the abstracts of journal articles, we now have access to the full texts of journal articles via PubMedCentral. Where research on clinical documents has been hampered by a lack of availability of data, we now have access to large bodies of data through the auspices of the Cincinnati Children's Hospital NLP Challenge, the i2b2 shared tasks (www.i2b2.org), and the new TREC Electronic Medical Records track, as well as gold standard data being generated under US-funded Strategic Health Advanced Research Projects Area 4 (www.sharpn.org). Meanwhile, the number of abstracts in PubMed continues to grow exponentially. These are exciting times for biomedical NLP.

The Biomedical Information Extraction Workshop at RANLP 2011 provides a venue for presentation of current work in this field. Submissions in the areas of medicine and biology were solicited, as was research on the genres of publications, clinical documents, and web-based materials.

One theme present in the workshop was medical coding, represented by the work in Henriksson and Hassel (2011) and Boytcheva (2011). Other papers dealt with parsing (Kokkinakis 2011), evaluation of detection of personal health information (Sokolova 2011), and named entity recognition (Georgiev and Nakov 2011). There was a clear trend in the workshop towards working with clinically oriented problems or data rather than publication data, exemplified by most of the preceding papers, as well as work by Proux et al. on hospital-acquired infections and by Angelova and Boytcheva (2011) on the difficult problem of temporality in discharge notes. This is encouraging for the field in terms of its implications for the growing availability of clinically oriented data. A final encouraging note is that work was presented on data other than English, making this workshop nearly unique in its provision of a venue for work on this important topic.

We would like to thank all the authors for their efforts in making the event a highly productive workshop and a lively venue for exchange of scientific ideas. We also thank the Programme Committee members and the reviewers for providing high quality reviews. The workshop is organised with the partial support of project D0-02-292 EVTIMA "Effective search of conceptual information with applications in medical informatics" funded by the Bulgarian National Science Fund in 2009-2012.

September 2011

Guergana Savova
Kevin Bretonnel Cohen
Galia Angelova

Organizers:

Guergana Savova (Children's Hospital Boston and Harvard Medical School)
Kevin Bretonnel Cohen (University of Colorado School of Medicine)
Galia Angelova (IICT Bulgarian Academy of Sciences)

Programme Committee:

Vangelis Karkaletsis (Institute of Informatics and Telecommunications, Athens)
Dimitris Kokkinakis (Gothenburg University, Norway)
Frédérique Segond (Xerox Research Centre Europe, Grenoble)
Preslav Nakov (National University of Singapore, Singapore)
Pinar Wennerberg (Bayer AG, Germany)

Additional Reviewers:

Svetla Boytcheva (State University of Library Studies and Information Technologies, Bulgaria)
Georgi Georgiev (OntoText AD, Bulgaria)
Jiaping Zheng (Children's Hospital Boston and Harvard Medical School)
Tim Miller (Children's Hospital Boston and Harvard Medical School)

Workshop Programme

Thursday, 15 September, 2011

9:30–10:30 *Invited talk: Assignment of Ontology-based Broad Semantic Classes to Biomedical Text*
Kevin Bretonnel Cohen

10:30–11:00 Coffee break

11:00–11:30 *Exploiting Structured Data, Negation Detection and SNOMED CT Terms in a Random Indexing Approach to Clinical Coding*
Aron Henriksson and Martin Hassel

11:30–12:00 *Automatic Matching of ICD-10 codes to Diagnoses in Discharge Letters*
Svetla Boytcheva

12:00–12:30 *Evaluation Measures for Detection of Personal Health Information*
Marina Sokolova

12:30–14:30 Lunch break

14:30-15:00 *Building a Named Entity Recognizer in Three Days: Application to Disease Name Recognition in Bulgarian Epicrises*
Georgi Georgiev, Valentin Zhivkov, Borislav Popov and Preslav Nakov

15:00–15:30 *Reducing Complexity in Parsing Scientific Medical Data, a Diabetes Case Study*
Dimitrios Kokkinakis

15:30–16:00 Coffee break

16:00–16:40 Poster Presentations

Architecture and Systems for Monitoring Hospital Acquired Infections inside Hospital Information Workflows
Denys Proux, Caroline Hagège, Quentin Gicquel, Suzanne Pereira, Stefan Darmoni, Frédérique Segond and Marie-Hélène Metzger

Towards Temporal Segmentation of Patient History in Discharge Letters
Galia Angelova and Svetla Boytcheva

Assignment of ontology-based broad semantic classes to biomedical text

K. Bretonnel Cohen
Computational Bioscience Program, U. Colorado School of Medicine
Department of Linguistics, U. Colorado at Boulder
kevin.cohen@gmail.com

Abstract

Natural language processing of biomedical text benefits from the ability to recognize broad semantic classes, but the number of semantic types is far bigger than is usually treated in newswire text. A method for broad semantic class assignment using lightweight linguistic analysis is described and evaluated using traditional and novel methods.

1 Introduction

Experience with coreference resolution, information extraction, and document classification indicates that these natural language processing tasks, and probably others, benefit from the ability to assign semantic classes to entities in text. However, the set of semantic classes in biomedical text is much larger than the traditional MUC categories of PERSON, ORGANIZATION, and LOCATION that are normally dealt with in newswire text. This talk addresses the hypothesis that it is possible to label as many as twenty semantic types in biomedical text, with those semantic classes being grounded in the topic areas of ontologies from the Open Biomedical Ontologies collection at the National Center for Biomedical Ontology. Belonging to the topic of an Open Biomedical Ontology is then considered to constitute membership in that semantic class. We determined membership in semantic classes through a lightweight linguistic analysis consisting of four techniques of decreasing levels of stringency.

2 Method

Twenty ontologies thought to be relevant to mouse genomics were selected from the Open Biomedical Ontologies collection. Four methods were used to match these to pre-extracted phrases in text. In the first technique, a simple exact match was attempted. In the second technique, a normalized form of the term, with whitespace and punctuation removed, was used to attempt a match. (Identical normalization was applied to the input text.)

In the third technique, a lightweight linguistic technique was applied. Simple heuristics were used to extract the head noun from each term in the ontology, and from the input text phrase. A match of the headwords was then attempted. If this failed, both headwords were stemmed and a match was attempted again.

The four techniques were applied sequentially. This allowed for modular evaluation of each technique, and at runtime, it allows for the user to select a minimum level of stringency.

The method was evaluated by three techniques:

1. Measuring precision, recall, F-measure, and accuracy with a unique corpus of full-text journal articles marked up with a number of biomedical ontologies.
2. Using the ontologies themselves as input. This is a novel evaluation technique, and it was shown to be robust in detecting errors in the method.
3. Running the method against a structured test suite (Cohen et al. 2010) used for ontology concept recognition.

Proceedings of the Workshop on Biomedical Natural Language Processing, pages 1–2,
Hissar, Bulgaria, 15 September 2011.

A micro-averaged F-measure of 72.32 was achieved on the corpus. The macro-averaged F-measure was 75.31. Accuracies of 77.12 to 95.73% were achieved, depending on the ontology. The lightweight linguistic technique of head noun extraction was found to make a significant contribution to efficacy, sometimes a very large contribution.

As would be expected, the evaluation against the ontologies themselves typically achieved very high performance numbers, but in two cases it uncovered significant lapses in our processing of the ontologies themselves, indicating that this novel evaluation method is robust.

Finally, the structured test set yielded considerable insight into the strengths and weaknesses of the method.

A full description of the materials, method, and evaluation can be found in Cohen et al. (2011).

References

Cohen, K. Bretonnel; Christophe Roeder; William A. Baumgartner Jr.; Lawrence E. Hunter; and Karin Verspoor (2010) Test suite design for biomedical ontology concept recognition systems. *Languages Resources and Evaluation Conference,* pp. 441-446.

Cohen, K. Bretonnel; Tom Christiansen; William A. Baumgartner Jr.; Karin Verspoor; and Lawrence E. Hunter (2011) Fast and simple semantic class assignment for biomedical text. *Biomedical Natural Language Processing 2011,* pp. 38-45.

Exploiting Structured Data, Negation Detection and SNOMED CT Terms in a Random Indexing Approach to Clinical Coding

Aron Henriksson
DSV, Stockholm University
aronhen@dsv.su.se

Martin Hassel
DSV, Stockholm University
xmartin@dsv.su.se

Abstract

The problem of providing effective computer support for clinical coding has been the target of many research efforts. A recently introduced approach, based on statistical data on co-occurrences of words in clinical notes and assigned diagnosis codes, is here developed further and improved upon. The ability of the word space model to detect and appropriately handle the function of negations is demonstrated to be important in accurately correlating words with diagnosis codes, although the data on which the model is trained needs to be sufficiently large. Moreover, weighting can be performed in various ways, for instance by giving additional weight to 'clinically significant' words or by filtering code candidates based on structured patient records data. The results demonstrate the usefulness of both weighting techniques, particularly the latter, yielding 27% exact matches for a general model (across clinic types); 43% and 82% for two domain-specific models (ear-nose-throat and rheumatology clinics).

1 Introduction

Clinicians spend much valuable time and effort in front of a computer, assigning diagnosis codes during or after a patient encounter. Tools that facilitate this task would allow costs to be reduced or clinicians to spend more of their time tending to patients, effectively improving the quality of healthcare. The idea, then, is that clinicians should be able simply to verify automatically assigned codes or to select appropriate codes from a list of recommendations.

1.1 Previous Work

There have been numerous attempts to provide clinical coding support, even if such tools are yet to be widely used in clinical practice (Stanfill et al., 2010). The most common approach has been to view it essentially as a text classification problem. The assumption is that there is some overlap between clinical notes and the content of assigned diagnosis codes, making it possible to predict possible diagnosis codes for 'uncoded' documents. For instance, in the *2007 Computational Challenge* (Pestian et al., 2007), free-text radiology reports were to be assigned one or two labels from a set of 45 ICD-9-CM codes. Most of the best-performing systems were rule-based, achieving micro-averaged F_1-scores of up to 89.1%.

Some have tried to enhance their NLP-based systems by exploiting the structured data available in patient records. Pakhomov et al. (2006) use gender information—as well as frequency data—to filter out improbable classifications. The motivation is that gender has a high predictive value, particularly as some categories make explicit gender distinctions.

Medical terms also have a high predictive value when it comes to classification of clinical notes (see e.g. Jarman and Berndt, 2010). In an attempt to assign ICD-9 codes to discharge summaries, the results improved when extra weight was given to words, phrases and structures that provided the most diagnostic evidence (Larkey and Croft, 1995).

Given the inherent practice of ruling out possible diseases, symptoms and findings, it seems important to handle negations in clinical text. In one study, it was shown that around 9% of automatically detected SNOMED CT *findings* and *disorders* were negated (Skeppstedt et al., 2011). In the attempt of Larkey and Croft (1995), negated medical terms are annotated and handled in various

Proceedings of the Workshop on Biomedical Natural Language Processing, pages 3–10,
Hissar, Bulgaria, 15 September 2011.

ways; however, none yielded improved results.

1.2 Random Indexing of Patient Records

In more recent studies, the *word space model*, in its *Random Indexing* mold (Sahlgren, 2001; Sahlgren, 2006), has been investigated as a possible alternative solution to clinical coding support (Henriksson et al., 2011; Henriksson and Hassel, 2011). Statistical data on co-occurrences of words and ICD-10[1] codes is used to build predictive models that can generate recommendations for uncoded documents. In a list of ten recommended codes, general models—trained and evaluated on all clinic types—achieve up to 23% exact matches and 60% partial matches, while domain-specific models—trained and evaluated on a particular type of clinic—achieve up to 59% exact matches and 93% partial matches.

A potential limitation of the above models is that they fail to capture the function of negations, which means that negated terms in the clinical notes will be positively correlated with the assigned diagnosis codes. In the context of information retrieval, Widdows (2003) describes a way to remove unwanted meanings from queries in vector models, using a vector negation operator that not only removes unwanted strings but also synonyms and neighbors of the negated terms. To our knowledge, however, the ability of the word space model to handle negations has not been studied extensively.

1.3 Aim

The aim of this paper, then, is to develop the Random Indexing approach to clinical coding support by exploring three potential improvements:

1. Giving extra weight to words used in a list of SNOMED CT terms.

2. Exploiting structured data in patient records to calculate the likelihood of code candidates.

3. Incorporating the use of negation detection.

2 Method

Random Indexing is applied on patient records to calculate co-occurrences of tokens (words and ICD-10 codes) on a document level. The resulting models contain information about the 'semantic similarity' of individual words and diagnosis codes[2], which is subsequently used to classify uncoded documents.

2.1 Stockholm EPR Corpus

The models are trained and evaluated on a Swedish corpus of approximately 270,000 clinically coded patient records, comprising 5.5 million notes from 838 clinical units. This is a subset of the *Stockholm EPR* corpus (Dalianis et al., 2009). A *document* contains all free-text entries concerning a single patient made on consecutive days at a single clinical unit. The documents in the partitions of the data sets on which the models are trained (90%) also include one or more associated ICD-10 codes (on average 1.7 and at most 47). In the testing partitions (10%), the associated codes are retained separately for evaluation. In addition to the complete data set, two subsets are created, in which there are documents exclusively from a particular type of clinic: one for *ear-nose-throat* clinics and one for *rheumatology* clinics.

Variants of the three data sets are created, in which negated clinical entities are automatically annotated using the Swedish version of *NegEx* (Skeppstedt, 2011). The clinical entities are detected through exact string matching against a list of 112,847 *SNOMED CT* terms belonging to the semantic categories 'finding' and 'disorder'. It is important to handle ambiguous terms in order to reduce the number of false positives; therefore, the list does not include findings which are equivalent to a common non-clinical unigram or bigram (see Skeppstedt et al., 2011). A negated term is marked in such a way that it will be treated as a single word, although with its proper negated denotation. Multi-word terms are concatenated into unigrams.

The data is finally pre-processed: lemmatization is performed using the *Granska Tagger* (Knutsson et al., 2003), while punctuation, digits and stop words are removed.

2.2 Word Space Models

Random Indexing is performed on the training partitions of the described data sets, resulting in

[1] The 10th revision of the *International Classification of Diseases and Related Health Problems* (World Health Organization, 2011).

[2] According to the *distributional hypothesis*, words that appear in similar contexts tend to have similar properties. If two words repeatedly co-occur, we can assume that they in some way refer to similar concepts (Harris, 1954). Diagnosis codes are here treated as words.

a total of six models (Table 1): two variants of the general model and two variants of the two domain-specific models[3].

Table 1: The six models.

w/o negations	w/ negations
General_Model	*General_NegEx_Model*
ENT_Model	*ENT_NegEx_Model*
Rheuma_Model	*Rheuma_NegEx_Model*

2.3 Election of Diagnosis Codes

The models are then used to produce a ranked list of recommended diagnosis codes for each of the documents in the testing partitions of the corresponding data sets. This list is created by letting each of the words in a document 'vote' for a number of semantically similar codes, thus necessitating the subsequent merging of the individual lists. This ranking procedure can be carried out in a number of ways, some of which are explored in this paper. The starting point, however, is to use the semantic similarity of a word and a diagnosis code—as defined by the cosine similarity score—and the idf[4] value of the word. This is regarded as our baseline model (Henriksson and Hassel, 2011), to which negation handling and additional weighting schemes are added.

2.4 Weighting Techniques

For each of the models, we apply two distinct weighting techniques. First, we assume a *technocratic* approach to the election of diagnosis codes. We do so by giving added weight to words which are 'clinically significant'. That is here achieved by utilizing the same list of SNOMED CT findings and disorders that was used by the negation detection system. However, rather than trying to match the entire term—which would likely result in a fairly limited number of hits—we opted simply to give weight to the individual (non stop) words used in those terms. These words are first lemmatized, as the data on which the matching is performed has also been lemmatized. It will also allow hits independent of morphological variations.

We also perform weighting of the correlated ICD-10 codes by exploiting statistics generated from the fixed fields of the patient records, namely *gender*, *age* and *clinical unit*. The idea is to use known information about a to-be-coded document in order to assign weights to code candidates according to plausibility, which in turn is based on past combinations of a particular code and each of the structured data entries. For instance, if the model generates a code that has very rarely been assigned to a patient of a particular sex or age group—and the document is from the record of such a patient—it seems sensible to give it less weight, effectively reducing the chances of that code being recommended. In order for an unseen combination not to be ruled out entirely, additive smoothing is performed. Gender and clinical unit can be used as defined, while age groups are created for each and every year up to the age of 10, after which ten-year intervals are used. This seems reasonable since age distinctions are more sensitive in younger years.

In order to make it possible for code candidates that are not present in any of the top-ten lists of the individual words to make it into the final top-ten list of a document, all codes associated with a word in the document are included in the final re-ranking phase. This way, codes that are more likely for a given patient are able to take the place of more improbable code candidates. For the general models, however, the initial word-based code lists are restricted to twenty, due to technical efficiency constraints.

2.5 Evaluation

The evaluation is carried out by comparing the model-generated recommendations with the clinically assigned codes in the data. This matching is done on all four possible levels of ICD-10 according to specificity (see Figure 1).

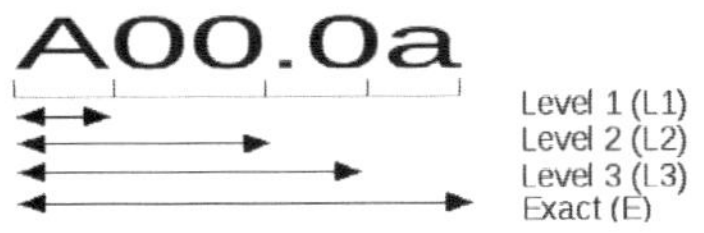

Figure 1: The structure of ICD-10 allows division into four levels.

3 Results

The general data set, on which *General_Model* and *General_NegEx_Model* are trained and evaluated, comprises approximately 274,000 documents and 12,396 unique labels. The ear-nose-throat data

[3]ENT = Ear-Nose-Throat, Rheuma = Rheumatology.

[4]Inverse document frequency, denoting a word's discriminatory value.

set, on which *ENT_Model* and *ENT_NegEx_Model* are trained and evaluated, contains around 23,000 documents and 1,713 unique labels. The rheumatology data set, on which *Rheuma_Model* and *Rheuma_NegEx_Model* are trained an evaluated, contains around 9,000 documents and 630 unique labels (Table 2).

data set	documents	codes
General	~274 k	12,396
ENT	~23 k	1,713
Rheumatology	~9 k	630

Table 2: Data set statistics.

The proportion of the detected clinical entities that are negated is 13.98% in the complete, general data set and slightly higher in the ENT (14.32%) and rheumatology data sets (16.98%) (Table 3).

3.1 General Models

The baseline for the general models finds 23% of the clinically assigned codes (exact matches), when the number of model-generated recommendations is confined to ten (Table 4). Meanwhile, matches on the less specific levels of ICD-10, i.e. partial matches, amount to 25%, 33% and 60% respectively (from specific to general).

The single application of one of the weighing techniques to the baseline model boosts performance somewhat, the fixed fields-based code filtering (26% exact matches) slightly more so than the technocratic word weighting (24% exact matches). The negation variant of the general model, *General_NegEx_Model*, performs somewhat better—up two percentage points (25% exact matches)—than the baseline model. The technocratic approach applied to this model does not yield any observable added value. The fixed fields filtering does, however, result in a further improvement on the three most specific levels (27% exact matches).

A combination of the two weighting schemes does not appear to bring much benefit to either of the general models, compared to solely performing fixed fields filtering.

3.2 Ear-Nose-Throat Models

The baseline for the ENT models finds 33% of the clinically assigned codes (exact matches) and 34% (L3), 41% (L2) and 62% (L1) at the less specific levels (Table 5).

Technocratic word weighing yields a modest improvement over the baseline model: one percentage point on each of the levels. Filtering code candidates based on fixed fields statistics, however, leads to a remarkable boost in results, from 33% to 43% exact matches. *ENT_NegEx_Model* performs slightly better than the baseline model, although only as little as a single percentage point (34% exact matches). Performance drops when the technocratic approach is applied to this model. The fixed fields filtering, on the other hand, similarly improves results for the negation variant of the ENT model; however, there is no apparent additional benefit in this case of negation handling. In fact, it somewhat hampers the improvement yielded by this weighting technique.

As with the general models, a combination of the two weighting techniques does not affect the results much for either of the ENT models.

3.3 Rheumatology Models

The baseline for the rheumatology models finds 61% of the clinically assigned codes (exact matches) and 61% (L3), 68% (L2) and 92% (L1) at the less specific levels (Table 6).

Compared to the above models, the technocratic approach is here much more successful, resulting in 72% exact matches. Filtering the code candidates based on fixed fields statistics leads to a further improvement of ten percentage points for exact matches (82%). *Rheuma_NegEx_Model* achieves only a modest improvement on L2. Moreover, this model does not benefit at all from the technocratic approach; neither is the fixed fields filtering quite as successful in this model (67% exact matches).

A combination of the two weighting schemes adds only a little to the two variants of the rheumatology model. Interesting to note is that the negation variant performs the same or even much worse than the one without any negation handling.

4 Discussion

The two weighting techniques and the incorporation of negation handling provide varying degrees of benefit—from small to important boosts in performance—depending to some extent on the model to which they are applied.

Model	Clinical Entities	Negations	Negations/Clinical Entities
General_NegEx_Model	634,371	88,679	13.98%
ENT_NegEx_Model	40,362	5,780	14.32%
Rheuma_NegEx_Model	20,649	3,506	16.98%

Table 3: *Negation Statistics*. The number of detected clinical entities, the number of negated clinical entities and the percentage of the detected clinical entities that are negated.

	General_Model				*General_NegEx_Model*			
Weighting	E	L3	L2	L1	E	L3	L2	L1
Baseline	*0.23*	*0.25*	*0.33*	*0.60*	0.25	0.27	0.35	0.62
Technocratic	0.24	0.26	0.34	0.61	0.25	0.27	0.35	0.62
Fixed Fields	0.26	0.28	0.36	0.61	0.27	0.29	0.37	0.63
Technocratic + Fixed Fields	0.26	0.28	0.36	0.62	0.27	0.29	0.37	0.63

Table 4: *General Models*, with and without negation handling. Recall (top 10), measured as the presence of the clinically assigned codes in a list of ten model-generated recommendations. E = exact match, L3→L1 = matches on the other levels, from specific to general. The baseline is for the model without negation handling only.

	ENT_Model				*ENT_NegEx_Model*			
Weighting	E	L3	L2	L1	E	L3	L2	L1
Baseline	*0.33*	*0.34*	*0.41*	*0.62*	0.34	0.35	0.42	0.62
Technocratic	0.34	0.35	0.42	0.63	0.33	0.33	0.41	0.61
Fixed Fields	0.43	0.43	0.48	0.64	0.42	0.43	0.48	0.63
Technocratic + Fixed Fields	0.42	0.42	0.47	0.64	0.42	0.42	0.47	0.62

Table 5: *Ear-Nose-Throat Models*, with and without negation handling. Recall (top 10), measured as the presence of the clinically assigned codes in a list of ten model-generated recommendations. E = exact match, L3→L1 = matches on the other levels, from specific to general. The baseline is for the model without negation handling only.

	Rheuma_Model				*Rheuma_NegEx_Model*			
Weighting	E	L3	L2	L1	E	L3	L2	L1
Baseline	*0.61*	*0.61*	*0.68*	*0.92*	0.61	0.61	0.70	0.92
Technocratic	0.72	0.72	0.77	0.94	0.60	0.60	0.70	0.91
Fixed Fields	0.82	0.82	0.85	0.95	0.67	0.67	0.75	0.91
Technocratic + Fixed Fields	0.82	0.83	0.86	0.95	0.68	0.68	0.76	0.92

Table 6: *Rheumatology Models*, with and without negation handling. Recall (top 10), measured as the presence of the clinically assigned codes in a list of ten model-generated recommendations. E = exact match, L3→L1 = matches on the other levels, from specific to general. The baseline is for the model without negation handling only.

4.1 Technocratic Approach

The technocratic approach, whereby clinically significant words are given extra weight, does result in some improvement when applied to all models that do not incorporate negation handling. The effect this weighting technique has on *Rheuma_Model* is, however, markedly different from when it is applied to the other two corresponding models. It could potentially be the result of a more precise, technical language used in rheumatology documentation, where certain words are highly predictive of the diagnosis. However, the results produced by this model need to be examined with some caution, due to the relatively small size of the data set on which the model is based and evaluated.

Since this approach appears to have a positive impact on all of the models where negation handling is not performed, assigning even more weight to clinical terminology may yield additional benefits. This would, of course, have to be tested empirically and may differ from domain to domain.

4.2 Structured Data Filtering

The technique whereby code candidates are given weight according to their likelihood of being accurately assigned to a particular patient record—based on historical co-occurrence statistics of diagnosis codes and, respectively, age, gender and clinical unit—is successful across the board. To a large extent, this is probably due to a set of ICD-10 codes being frequently assigned in any particular clinical unit. In effect, it can partly be seen as a weighting scheme according to code frequency. There are also codes, however, that make gender and age distinctions. It is likewise well known that some diagnoses are more prevalent in certain age groups, while others are exclusive to a particular gender.

It is interesting to note the remarkable improvement observed for the two domains-specific models. Perhaps the aforementioned factor of frequently recurring code assignments is even stronger in these particular types of clinics. By contrast, there are no obvious gender-specific diagnoses in either of the two domains; however, in the rheumatology data, there are in fact 23 codes that have frequently been assigned to men but never to women. In such cases it is especially beneficial to exploit the structured data in patient records. It could also be that the restriction to twenty code candidates for each of the individual words in the general models was not sufficiently large a number to allow more likely code candidates to make it into the final list of recommendations. That said, it seems somewhat unlikely that a code that is not closely associated with any of the words in a document should make it into the final list.

Even if the larger improvements observed for the domain-specific models may, again, in part be due to the smaller amounts of data compared with the general model, the results clearly indicate the general applicability and benefit of such a weighting scheme.

4.3 Negation Detection

The incorporation of automatic detection of negated clinical entities improves results for all models, although more so for the general model than the domain-specific models. This could possibly be ascribed to the problem of data sparsity. That is, in the smaller domain-specific models, there are fewer instances of each type of negated clinical entity (11.7 on average in ENT and 9.4 on average in rheumatology) than in the general model (31.6 on average). This is problematic since infrequent words, just as very frequent words, are commonly assumed to hold little or no information about semantics (Jurafsky and Martin, 2009). There simply is little statistical evidence for the rare words, which potentially makes the estimation of their similarity with other words uncertain. For instance, Karlgren and Sahlgren (2001) report that, in their TOEFL test experiments, they achieved the best results when they removed words that appeared in only one or two documents. While we cannot just remove infrequent codes, the precision of these suggestions are likely to be lower.

The prevalence of negated clinical entities—almost 14% in the entire data set—indicates the importance of treating them as such in an NLP-based approach to clinical coding. Due to the extremely low recall (0.13) of the simple method of detecting clinical entities through exact string matching (Skeppstedt et al., 2011), negation handling could potentially have a more marked impact on the models if more clinical entities were to be detected, as that would likely also entail more negated terms.

There are, of course, various ways in which one may choose to handle negations. An alternative could have been simply to ignore negated terms in the construction of the word space models, thereby not correlating negated terms with affirmed diagnosis codes. Even if doing so may make sense, the approach assumed here is arguably better since a negated clinical entity could have a positive correlation with a diagnosis code. That is, ruling out or disconfirming a particular diagnosis may be indicative of another diagnosis.

4.4 Combinations of Techniques

When the technocratic weighting technique is applied to the variants of the models which include annotations of negated clinical entities, there is no positive effect. In fact, results drop somewhat when applied to the two domain-specific models. A possible explanation could perhaps be that clinically significant words that are constituents of negated clinical entities are not detected in the technocratic approach. The reason for this is that the application of the Swedish NegEx system, which is done prior to the construction and evaluation of the models, marks the negated clinical entities in such a way that those words will no longer be recognized by the technocratic word detector. Such words may, of course, be of importance even if they are negated. This could be worked around in various ways; one would be simply to give weight to all negated clinical entities.

Fixed fields filtering applied to the NegEx models has an impact that is more or less comparable to the same technique applied to the models without negation handling. This weighting technique is thus not obviously impeded by the annotations of negated clinical entities, with the exception of the rheumatology models, where an improvement is observed, yet not as substantial as when applied to *Rheuma_Model*.

A combination of the technocratic word weighting and the fixed fields code filtering does not appear to provide any added value over the sole application of the latter weighting technique. Likewise, the same combination applied to the NegEx version does not improve on the results of the fixed fields filtering.

In this study, fine-tuning of weights has not been performed, neither internally or externally to each of the weighting techniques. It may, of course, be that, for instance, gender distinctions are more informative than age distinctions—or vice versa—and thus need to be weighted accordingly. By the same token should the more successful weighting schemes probably take precedence over the less successful variants.

4.5 Classification Problem

It should be pointed out that the model-generated recommendations are restricted to a set of properly formatted ICD-10 codes. Given the conditions under which real, clinically generated data is produced, there is bound to be some noise, not least in the form of inaccurately assigned and ill-formatted diagnosis codes. In fact, only 67.9% of the codes in the general data set are in this sense 'valid' (86.5% in the ENT data set and 66.9% in the rheumatology data set). As a result, a large portion of the assigned codes in the testing partition cannot be recommended by the models, possibly having a substantial negative influence on the evaluation scores. For instance, in the ear-nose-throat data, the five most frequent diagnosis codes are not present in the restricted result set. Not all of these are actually 'invalid' codes but rather action codes etc. that were not included in the list of acceptable code recommendations. A fairer evaluation of the models would be either to include such codes in the restricted result set or to base the restricted result set entirely on the codes in the data. Furthermore, there is a large number of unseen codes in the testing partitions, which also cannot be recommended by the models (358 in the general data set, 79 in the ENT data set and 39 in the rheumatology data set). This, on the other hand, reflects the real-life conditions of a classification system and so should not be eschewed; however, it is interesting to highlight when evaluating the successfulness of the models and the method at large.

5 Conclusion

The Random Indexing approach to clinical coding benefits from the incorporation of negation handling and various weighting schemes. While assigning additional weight to clinically significant words yields a fairly modest improvement, filtering code candidates based on structured patient records data leads to important boosts in performance for general and domain-specific models alike. Negation handling is also important, although the way in which it is here performed seems to require a large amount of training data for marked benefits. Even if combining a number of weighting techniques does not necessarily give rise to additional improvements, tuning of the weighting factors may help to do so.

Acknowledgments

We would like to thank all members of our research group, *IT for Health*, for their support and input. We would especially like to express our gratitude to Maria Skeppstedt for her important contribution to the negation handling aspects of this work, particularly in adapting her Swedish NegEx system to our specific needs. Thanks also to the reviewers for their helpful comments.

References

Hercules Dalianis, Martin Hassel and Sumithra Velupillai. 2009. The Stockholm EPR Corpus: Characteristics and Some Initial Findings. In Proceedings of ISHIMR 2009, pp. 243–249.

Zellig S. Harris. 1954. Distributional structure. *Word*, 10, pp. 146–162.

Aron Henriksson, Martin Hassel and Maria Kvist. 2011. Diagnosis Code Assignment Support Using Random Indexing of Patient Records — A Qualitative Feasibility Study. In Proceedings of AIME, 13th Conference on Artificial Intelligence in Medicine, pp. 348–352.

Aron Henriksson and Martin Hassel. 2011. Election of Diagnosis Codes: Words as Responsible Citizens. In Proceedings of Louhi, 3rd International Workshop on Health Document Text Mining and Information Analysis.

Jay Jarman and Donald J. Berndt. 2010. Throw the Bath Water Out, Keep the Baby: Keeping Medically-Relevant Terms for Text Mining. In Proceedings of AMIA, pp. 336–340.

Daniel Jurafsky and James H. Martin. 2009. *Speech and Language Processing. An Introduction to Natural Language Processing, Computational Linguistics, and Speech Recognition.* Pearson Education International, NJ, USA, p. 806.

Jussi Karlgren and Magnus Sahlgren. 2001. From Words to Understanding. *Foundations of Real-World Intelligence*, pp. 294–308.

Ola Knutsson, Johnny Bigert and Viggo Kann. 2003. A Robust Shallow Parser for Swedish. In Proceedings of Nodalida.

Leah S. Larkey and W. Bruce Croft. 1995. Automatic Assignment of ICD9 Codes to Discharge Summaries. In PhD thesis University of Massachusetts at Amherst, Amerst, MA, USA.

Serguei V.S. Pakhomov, James D. Buntrock and Christopher G. Chute. 2006. Automating the Assignment of Diagnosis Codes to Patient Encounters Using Example-based and Machine Learning Techniques. *J Am Med Inform Assoc*, 13, pp. 516–525.

John P. Pestian, Christopher Brew, Pawel Matykiewicz, DJ Hovermale, Neil Johnson, K. Bretonnel Cohen and Wlodzislaw Duch. 2007. A Shared Task Involving Mulit-label Classification of Clinical Free Text. In Proceedings of BioNLP 2007: Biological, translational, and clinical language processing, pp. 97–104.

Magnus Sahlgren. 2001. Vector-Based Semantic Analysis: Representing Word Meanings Based on Random Labels. In Proceedings of Semantic Knowledge Acquisition and Categorization Workshop at ESSLLI'01.

Magnus Sahlgren. 2006. The Word-Space Model: Using distributional analysis to represent syntagmatic and paradigmatic relations between words in high-dimensional vector spaces. In PhD thesis Stockholm University, Stockholm, Sweden.

Maria Skeppstedt. 2011. Negation detection in Swedish clinical text: An adaption of Negex to Swedish. *Journal of Biomedical Semantics 2*, S3.

Maria Skeppstedt, Hercules Dalianis and Gunnar H. Nilsson. 2011. Retrieving disorders and findings: Results using SNOMED CT and NegEx adapted for Swedish. In Proceedings of Louhi, 3rd International Workshop on Health Document Text Mining and Information Analysis.

Mary H. Stanfill, Margaret Williams, Susan H. Fenton, Robert A. Jenders and William R Hersh. 2010. A systematic literature review of automated clinical coding and classification systems. *J Am Med Infrom Assoc*, 17, pp. 646–651.

Dominic Widdows. 2003. Orthogonal Negation in Vector Spaces for Modelling Word-Meanings and Document Retrieval. In Proceedings of ACL, pp. 136-143.

World Health Organization. 2011. International Classification of Diseases (ICD). In *World Health Organization*. Retrieved June 19, 2011, from http://www.who.int/classifications/icd/en/.

Automatic Matching of ICD-10 codes to Diagnoses in Discharge Letters

Svetla Boytcheva
State University of Library Studies and Information Technologies,
119, Tzarigradsko Shosse, 1784 Sofia, Bulgaria
svetla.boytcheva@gmail.com

Abstract

This paper presents an approach for automatic mapping of International Classification of Diseases 10th revision (ICD-10) codes to diagnoses extracted from discharge letters. The proposed algorithm is designed for processing free text documents in Bulgarian language. Diseases are often described in the medical patient records as free text using terminology, phrases and paraphrases which differ significantly from those used in ICD-10 classification. In this way the task of diseases recognition (which practically means e.g. assigning standardized ICD codes to diseases' names) is an important natural language processing (NLP) challenge. The approach is based on multiclass Support Vector Machines method, where each ICD-10 4 character classification code is considered as single class. The problem is reduced to multiple binary classifiers and classification is done by a max-wins voting strategy.

1 Introduction

The nomenclature ICD or ICD CM (International Classification of Diseases with Clinical Modification), supported by WHO (World Health Organization) [1], is translated to many languages and serves as the main source for diagnoses definition.

The Bulgarian hospitals are reimbursed by the National Insurance Fund via the "clinical pathways" scheme. When a patient is hospitalized, they often select from the Hospital Information System (HIS) menu one diagnosis which is sufficient for the association of the desired clinical pathway to the respective patient. Thus most of complementary diseases diagnosed by the medical experts are recorded in the personal history as free text. To describe diseases as free text in the medical patient records (PRs) usually is used different terminology than those used in ICD-10 classification in order to express more specific and detailed information concerning particular disorder or using paraphrases, which usually are not available in general classification. For instance, in some diagnoses is specified the stage "затлъстяване I степен" (Stage 1 Obesity), the specific location "катаракта на ляво око" (left eye cataract) etc.

Thus the task of diagnoses recognition from free-text discharge letters and assignment of standardized ICD codes to diseases' names is an important natural language processing (NLP) challenge [2].

PRs in all Bulgarian hospitals have mandatory structure, which is published in the Official State Gazette within the legal Agreement between the Bulgarian Medical Association and the National Health Insurance Fund [3]. PRs contain the following sections: *(i)* personal data; *(ii)* diagnoses; *(iii)* anamnesis; *(iv)* patient status; *(v)* lab data; *(vi)* medical examiners comments; *(vii)* discussion; *(viii)* treatment; and *(ix)* recommendations. Most of the diagnoses are entered in the discharge letter section *Diagnoses* as free text and some of them are only mentioned in the *Discussion* or *Medical examiners comments*.

In this paper we present an approach based on multiclass Support Vector Machines (SVM) for automatic diagnoses recognition from free-text PRs and assignment the ICD-10 codes to them.

The paper is organized as follows: Section 2 overviews related work, Section 3 describes recourse bank, Section 4 presents the method, system architecture and some examples, Section 5

Proceedings of the Workshop on Biomedical Natural Language Processing, pages 11–18,
Hissar, Bulgaria, 15 September 2011.

discusses evaluation and results and Section 6 sketches further work and conclusion.

2 Related Work

The application of natural language processing methods to clinical free-text is of growing interest for both health care practitioners and academic researchers. Unfortunately there is no international standard for discharge letters presentation. Another main difficulty in such texts is medical terminology - German, English and French medical terminology mainly is based on domesticated terms, but still some Latin terms are used and some of them are modified by preserving Latin root and using domestic ending

There are no systems dealing with clinical texts in Bulgarian. Thus we will overview some of the recent results achieved mainly for processing discharge letters in English [6, 8, 9], German [7] and French.

Several methods dealing with this problem were presented on 2007 Computational Medicine Challenge where about 50 participants submitted results [4]. The main goal was to create and train computational intelligence algorithms that automate the assignment of ICD-9-CM codes to anonymised radiology reports with a training set of 978 documents and a test set of 976 documents.

In NLP the performance accuracy of text extraction procedures usually is measured by the *precision* (percentage of correctly extracted entities as a subset of all extracted entities), *recall* (percentage correctly extracted entities as a subset of all entities available in the corpus) and their harmonic mean

F-measure: F=2*Precision*Recall/(Precision+Recall).

In 2007 Computational Medicine Challenge [5] the top-performing systems achieved F-measure 0.8908, the minimum was with F-measure 0.1541, and the mean was 0.7670, with a standard deviation of 0.1340. Some 21 systems have F-measure between 0.81 and 0.90. Another 14 systems have F-measure 0.70. The article [9] compares three machine learning methods on radiological reports and points out that the best F-measure is 77%.

The top rated systems use variety of approaches like: Machine-learning; Symbolic methods; Hybrid approaches; UMLS Structures, Robust classification algorithm (naive Bayes) etc. the system reported in [8] uses a hybrid approach combining example-based classification and a simple but robust classification algorithm (naive Bayes) with high performance over 22 million PRs: F-measure 98,2%; for about 48% of the medical records at Mayo clinic. SynDiKATe [7] based on combination between text parsing and semantic information derivation from a Bayesian network and reports about 76% F-measure.

The better systems process negations, hypernyms and synonyms and were apparently doing significant amounts of symbolic processing.

SVMs and related approaches to machine learning were strongly represented in this challenge, but did not seem to be reliably predictive of high ranking. This motivated us to try to use SVM method for ICD-10 codes assignment to diagnoses, but enhanced with some preprocessing techniques applied to input data, concerning usage of synonyms, hyponyms, negation processing and word normalization and etc. used in other methods with better performance.

3 Material

The IE experiments were performed on training corpus if 1,300 and test corpus of 6200 anonymised hospital PRs for patients with endocrine and metabolic diseases provided by the University Specialised Hospital for Active Treatment of Endocrinology (USHATE), Medical University Sofia, Bulgaria.

Bulgarian medical texts contain a specific mixture of terminology in Latin, Bulgarian and Latin terms transcribed with Cyrillic letters (Table 1). There is no preferred language for the terminology so the two forms are used like synonyms. The terms occur in the text with a variety of wordforms which is typical for the highly-inflexional Bulgarian language.

The mixture of such terminology, given in Cyrillic and Latin alphabets, makes very hard the task for automatic assignment of ICD-10 codes to diagnoses. About 2.34% of the text is presented with Latin letters; the rest is written with Cyrillic symbols but contains Latin terminology (mostly diagnoses, anatomic organs and examinations) which is transliterated to Cyrillic alphabet (about 11.6% of all terms). About 37% of all diagnoses in our test corpus of 1,300 PRs were presented in Latin. This very specific medical language reflects the established medical tradition to use Latin language. Last but not least the foreign terminology is due to the lack of controlled vocabularies in Bulgarian language. In addition no bilingual Bulgarian-Latin medical dictionary is available in electronic format as well.

Table 1 Examples for diagnoses representation

Type	Example
Mixture of medical terminology in Latin and Bulgarian	Консултация с офталмолог: ВОД= 0,6 ВОС=0,6, двуочно 0,8 с корекция. Фундоскопия: папили на нивото на ретината. Angiosclerosis vas. retinae hypertonica. Начални промени по типа на диабетна ретинопатия.
Medical terminology in Bulgarian	Диагноза: Захарен диабет тип 2. Затлъстяване II ст. Диабетна полиневропатия. Артериална хипертония-IIст. Диенцефален синдром.
Latin terms transcribed with Cyrillic letters	Диагноза: Хипопаратиреоидизмус постоператива компенсата. Хипотиреоидизмус Постоператива компенсата. Статус пост тиреоидектомиам про карцинома папиларе лоби синистри. Статус пост радиойодаблациам.

Further we have developed semi-automatically a dictionary with pairs of Latin and Bulgarian terms corresponding to anatomic organs and their status containing about 7,230 terms. The most complicated task was to develop semi-automatically the list of correspondences between diagnoses in Bulgarian and Latin. For this task were used resources available in Bulgarian [10] and English [11]: ICD-10 Classification (Fig. 1); Index of diseases and pathological states and their modifications (Fig. 2). There were also used: *(i)* Terminologia Anatomica providing terminology in English and Latin; *(ii)* Sets of about 300 prefixes and suffixes, about 100 roots, about 150 abbreviations in Latin and Greek and their corresponding meanings in English and Bulgarian; *(iii)* Rules for transliteration from Latin to Cyrillic.

E00.9	Вроден йод-недоимъчен синдром, неуточнен
E01.0	Дифузна (ендемична) гуша, свързана с йоден недоимък
E01.1	Полинодозна (ендемична) гуша, свързана с йоден недоимък
E01.2	Гуша (ендемична), свързана с йоден недоимък, неуточнена
E01.8	Други болести на щитовидната жлеза, свързани с йоден недоимък и сродни състояни
E03.0	Вроден хипотиреоидизъм с дифузна гуша
E03.1	Вроден хипотиреоидизъм без гуша
E03.2	Хипотиреоидизъм, дължащ се на лекарства и други екзогенни вещества
E03.3	Постинфекциозен хипотиреоидизъм
E03.4	Атрофия на щитовидната жлеза (придобита)
E03.5	Микседемна кома
E03.8	Други уточнени видове хипотиреоидизъм
E03.9	Хипотиреоидизъм, неуточнен
E04.0	Нетоксична дифузна гуша
E04.1	Нетоксичен единичен възел на щитовидната жлеза
E04.2	Нетоксична полинодозна гуша
E04.8	Други уточнени видове нетоксична гуша

Fig. 1 ICD-10 Classification in Bulgarian - excerpt for class "E"

Коронавирус (coronavirus), като причина за болест, класифицирана другаде B97.2
Коронарна (артерия) – *виж състояние*
Корсиканска треска (*виж също* Малария) B54
Кортико-адренален – *виж състояние*
Коса – *виж също състояние*
– изтръгване, скубане на F63.3
– – разстройство в стереотипа на движенията F98.4
Косвен – *виж състояние*
Косми
– враснали L73.1
– пръстеновидни или извити (вродени) Q84.1

Криза
– адисонова E27.2
– бъбречна N28.8
– вторична глаукома H40.4
– емоционална F43.2
– – остра реакция на стрес F43.0
– – приспособителна реакция F43.2
– – специфична за детството и юношеството F93.8
– коремна R10.4
– на Pel (табесна) A52.1
– надбъбречна (корова) E27.2
– нитритна
– – адекватно назначено и правилно приложено вещество I95.2
– – свръхдоза или погрешно вещество,

Fig. 2 ICD-10 Volume 2 Tabular Index in Bulgarian – excerpt for "K" terms

ICD-10-CM (Clinical Modification) codes may consist of up to seven digits (Fig 3). A seventh character is required on some diagnoses that begin with "M," "O," "R," "S," "T," and "VWXY." and represents visit encounter or squeal for injuries and external causes.

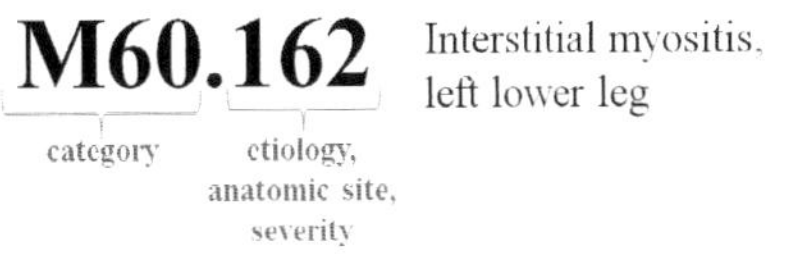

Fig. 3 ICD-10-CM Code Format

The ICD-10-CM is divided into the Alphabetic Index, an alphabetical list of terms and their corresponding code, and the Tabular List, a chronological list of codes divided into chapters based on body system or condition. The Alphabetic Index consists of the following parts: the Index of Diseases (Fig. 2) and Injury, the Index of External Causes of Injury, the Table of Neoplasms (Fig. 4) and the Table of Drugs and Chemicals [10].

	Злокачествено				
	Първично	Вторично	In situ	Доброкачествено	С неопределен или неизвестен характер
Новообразуване—*продължение*					
– външен(-на)					
– – маточна ос	C53.1	C79.8	D06.1	D26.0	D39.0
– – отвор (ухо)	C44.2	C79.2	D04.2	D23.2	D48.5
– вътрегръдна (кухина) (органи, НКД)	C76.1	C79.8		D36.7	D48.7
– вътреочно	C69.4	C79.4	D09.2	D31.4	D48.7
– вътрешна					
– – капсула	C71.0	C79.3		D33.0	D43.0
– – ос	C53.0	C79.8	D06.0	D26.0	D39.0
– вътрешни органи, НКД	C76.7	C79.8		D36.7	D48.7

Fig. 4 ICD-10 Volume 2 Table of Neoplasm

The data from ICD-10 Volume 2 Tabular index [10] are organized with leading term – level 1 (diagnose or pathological state) and modifications, which can be specified up to 7 levels. For instance for "A" terms are used 18256 words in total for explanation in different levels and 3568 different words.

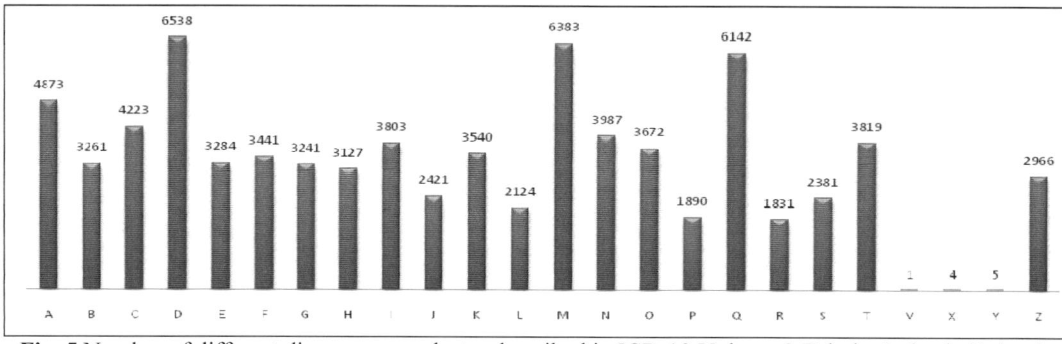

Fig. 5 Number of different diagnoses per cluster described in ICD-10 Volume 2 Tabular Index in Bulgarian

Searching in different sublevels not necessary specifies the ICD-10 codes. For instance, if the leading term is “Cyst” [10, 11], modification on level 1 “development” leads to K09.1, but further modification on level 2 “ovary” or “ovarian” leads to codes Q50.1 which belongs to other cluster. Another example for “Cyst” with modification on level 1 “epidermal” leads to L72.0 and further modification on level 2 “mouth” or “oral soft tissue” leads to codes K09.8. This shows that we need to use all nested levels of modifiers before final conclusion for the correct ICD-10 code for some diagnose.

In addition Tabular Index contains 19,161 different words and 291,116 words in total with repetitions, 2,221 in Latin (11.59%) and occurrences 83,713 in total (about 28.76%). This shows that direct application of Tabular index is not appropriate for automatic ICD-10 codes association for free-text diagnoses from discharge letters.

Tabular Index contains 76,939 descriptions of ICD-10 codes representing 9,044 different codes. ICD-10 classification [10] contains 14,439 different codes descriptions.

4 Method

4.1 SVM Classifier

SVMs have an ability to learn independent of the dimensionality of the feature space. This makes SVM Classifiers suitable approach for our task for automatic assignment of ICD-10 codes to diagnoses extracted from free-text PRs. This means that we can generalize even in the presence of very many features, because SVMs use overfitting protection, which does not necessarily depends on the number of features [12].

We present all ICD-10 codes as a set $C = \{c_1, c_2, \ldots, c_k\}$. The distribution in different clusters of codes described in the Tabular index (Fig. 5) shows that most of the diagnoses, except "VWXY" clusters, are described with variety of descriptions that in our opinion should be enough for generating rules for automatic classification.

To cover all possible codes included in ICD-10 classification we create training set of pairs $\left(x_i, c_j\right)$ of diagnoses descriptions x_i and their corresponding ICD-10 codes $c_j \in C$ from extracted diagnoses descriptions from Tabular Index in Bulgarian, those used in ICD-10 classification and 1,300 PRs from training corpus. In the training set vectors x_i contains words used to describe diagnose with omitting meaningless word (e.g. a, an, the, this, that, and, or).

The implemented system (Fig. 6) works in two steps [13,14]: (i) Preprocessing and (ii) SVM Classification.

Preprocessing analysis includes several text processing tasks performed as pipeline: PRs sections splitting; Tokenization; Diagnoses extraction; Abbreviations expansion; Transliteration; Latin terminology processing; Words normalization; Medical terminology synonyms; SVM model.

Bulgarian hospitals discharge letters have mandatory structure [3]. The system splits the text on all available sections and passes *Diagnose* section text for further processing. *Diagnose* section text is spitted into words set $W = \{w_1, w_2, \ldots, w_p\}$. Using scoping rules applied to *Diagnose* section text words from the generated set W are combined into diagnoses $D = \{d_1, d_2, \ldots, d_n\}$. For each diagnose $d_m \in D$ we create vector $y_{mi} =< w_{m1}, w_{m2}, \ldots, w_{mq} >$ containing words included in it. Using sets AL and AB of abbreviations in Latin and Bulgarian language and functions $a_l : AL \to Lat$ and

$a_b : AB \rightarrow Bul$, words in vectors y_{mi} for each diagnose $d_m \in D$ are substituted by expanded terms meaning in Latin (Lat) and Bulgarian (Bul) language correspondingly according (1).

$$u_{mj} = \begin{cases} a_b(w_{mj}), \text{ if } w_{mj} \in AB \\ a_l(w_{mj}), \text{if } w_{mj} \in AL \\ w_{mj} \quad , \text{ otherwise} \end{cases} \quad (1)$$

Then we replace vectors y_{mi} by their corresponding vectors $z_{mi} =< u_{m1}, u_{m2}, \ldots, u_{mq} >$ for each diagnose $d_m \in D$. Using transliteration rules $t : Cyrillic \rightarrow Latin$ from Cyrillic to Latin alphabet we convert each word in vector z_{mi} to its equivalent in Latin. The cases when some words in vector z_{mi} are in Latin and the other are in Cyrillic are very rare. If $t(u_{mj}) \in Lat$ or $u_{mj} \in Lat$ we substitute it by its corresponding term $b_{mj} \in Bul$ from Bulgarian terminology repository *Bul*, otherwise we suppose that u_{mj} is in Bulgarian and set $b_{mj} = u_{mj}$. Using rules for words derivatives we replace all terms b_{mj} by their lemmas $l_{mj} \in Bul$ and construct vector $v_{mi} =< l_{m1}, l_{m2}, \ldots, l_{mr} >$. The result vector v_{mi} contains only words in Bulgarian. Further we process negations [15] and searching for synonyms and hyponyms of disease and pathological states names in Bulgarian medical terminology repository. We generate all possible partitions P_{mj} of consecutive words in vector v_{mi}:

$$P_{mj} = \{l_{m1}, \ldots, l_{mq} \mid l_{mq+1}, \ldots, l_{mt} \mid \ldots \mid l_{ms}, \ldots, l_{mr}\} \quad (2)$$

For each sequence in (2) in partition P_{mj} we create set of its synonyms s_l. Usually parts contain from 1 up to 7 consecutive words in the vector. The Cartesian product of synonym sets for partition P_{mj} generates set $Y_{mj} = s_1 \times \ldots \times s_q = \{y_1, \ldots, y_p\}$ of input vectors. The union (3) of these sets for all partitions contains input vectors with different descriptions of each diagnose d_m:

$$Y_m = \bigcup_{P_{mj}} Y_{mj} \quad (3)$$

We use the formal representation for SVM model, learned by training examples, to transform test examples as input vectors for SVM.

SVM Classification - The input space in SVM Classifier is a vector space and the output is a single number corresponding to different classes. **SVM** classifier applies binary classification for each of the input vectors $y_t \in Y_m$ for each diagnose $d_m \in D$ and each of the classes in C. Winning strategy ranks all classes and chooses the highest ranked class c_{im} for each diagnose $d_m \in D$.

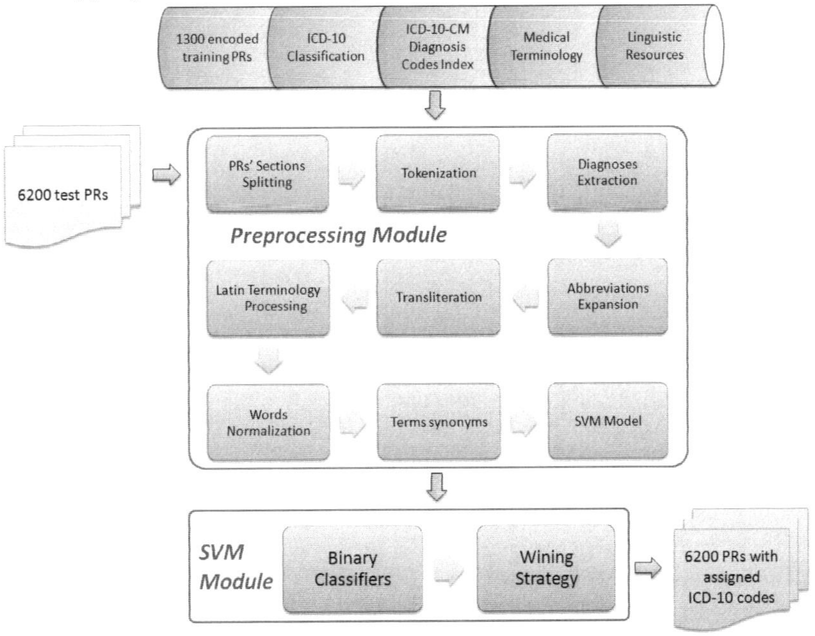

Fig. 6 System Architecture

4.2 Example

The implemented system allows processing of a single PR stored as text file in manual mode. There is also available automatic mode where can be processed all PRs stored in the selected folder and the result is stored in single CVS format file. In manual mode (Fig. 7) the text of PR is opened in section (1). After opening the text file, PR first is applied preprocessing steps from the algorithm and PR is automatically separated on sections and the text from diagnoses section is displayed in section (2). After choosing "Analyze" function from menu bar the extracted text in section (2) is processed and automatically is generated list (3) with recognized diagnoses within the text.

After selection of diagnose from list (3) to be processed its name is automatically excluded from list (3) and displayed in section (5). In the current example the selected diagnose *"киста оварии декстра" (киста на яйчника – in Bulgarian, cyst of ovary – in English)* is displayed in sections (5). The system identifies possible ICD-10 codes assignments and displays them in list (4) - *N83.0 Фоликуларна киста на яйчника (N83.0 Follicular cyst of ovary)*. It is possible the system to identify more than one possible codes for assignment, in this case different options are displayed in list (4) in decreasing order of ranking. The most appropriate association is ranked first. The data for processed diagnoses from list (3) are displayed in list (6) for further storage in CVS format text file.

In this example the diagnose *"феохромоцитома" (pheochromocytoma)* is presented using Latin terminology with transliteration. This term corresponds to *"Доброкачествено новообразувание на надбъбречна жлеза" (neoplasm of Adrenal gland)* in Bulgarian language. In ICD-10 4 chars categories it corresponds to D35.0. The next diagnose *"киста оварии декстра" (киста на яйчника – in Bulgarian, cyst of ovary – in English)* is processed using again latin terminology transliteration for Latin term *"оварии" (ovarian) (яйчници – in Bulgarian)* and the result assigned code is *N83.0 Фоликуларна киста на яйчника (N83.0 Follicular cyst of ovary)*. Here *"декстра"* in Latin means (*дясна – in Bulgarian, Right – in English*) is not considered in classification in this case.

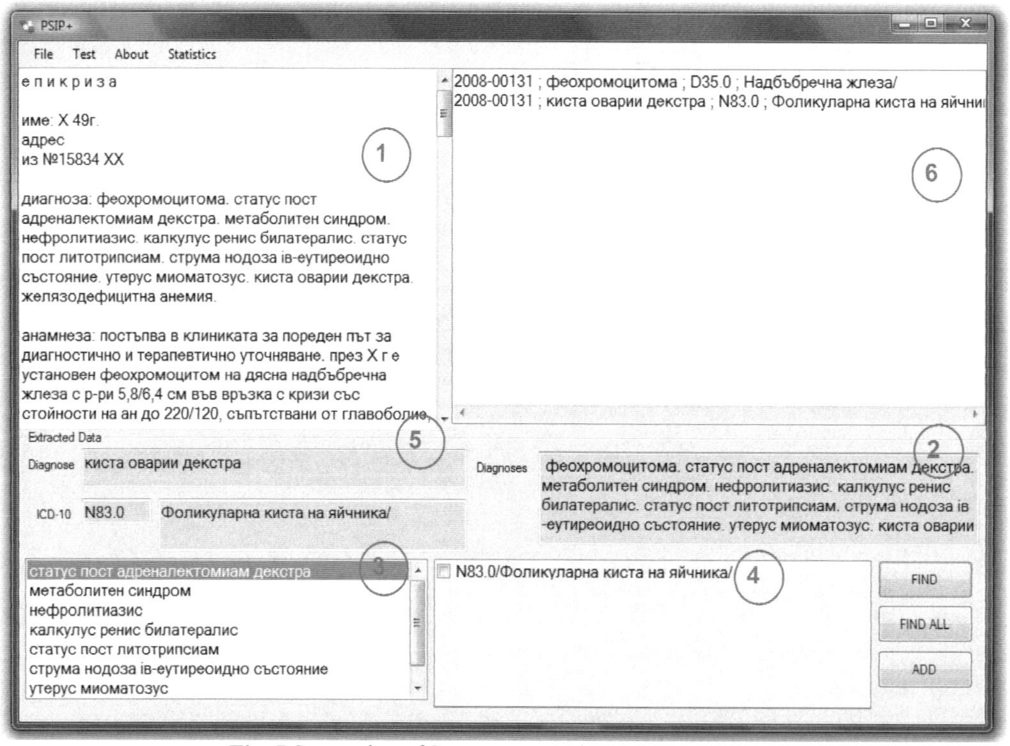

Fig. 7 Screenshot of System processing PR in manual mode

5 Evaluation and Results

The experiments were made with a training corpus described in Section 3 and the evaluation results are obtained using a test corpus, containing 6,200 PRs for patients with endocrine and metabolic diseases provided by USHATE.

For the test corpus there was identified descriptions of 26,826 diagnoses and 448 different classes diagnoses.

Because for the purposes of our project we are processing PRs for patients with endocrine and metabolic diseases their leading diagnoses are obviously classified in cluster "E". Thus some of the clusters are presented by few classified diagnoses (Fig. 8) and the other clusters (K, M, N, H, D, G, I) representing endocrine and metabolic diseases and related to them complications are presented by several classifications (Fig. 9).

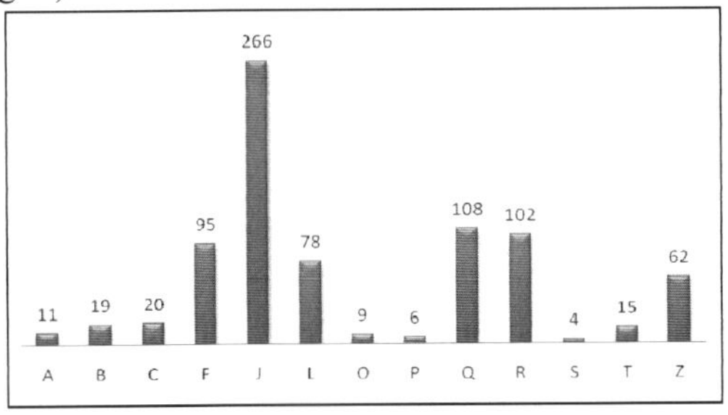

Fig. 8 Number of diagnoses classified for rare clusters

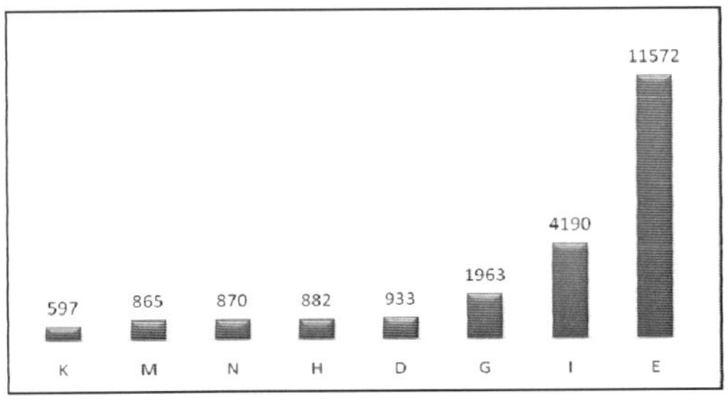

Fig. 9 Number of diagnoses classified for most common clusters

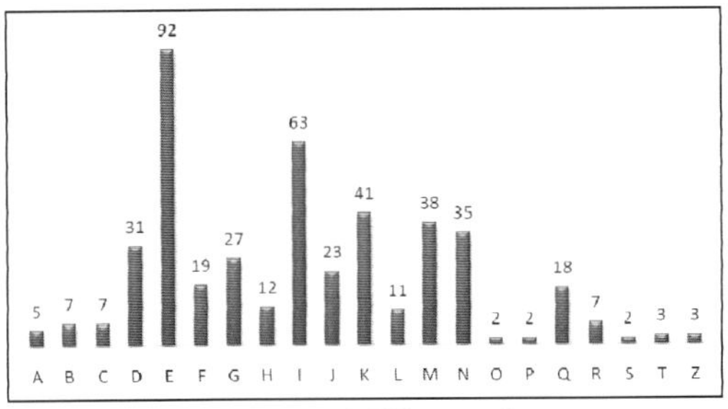

Fig. 10 Number of different diagnoses per cluster

Although the experiments was performed for such specific test set the diversity of result classes of diagnoses in test set presented on Fig. 10 and the average number of classified diagnoses per cluster (Fig. 11) shows that almost all cluster were presented by sufficient amount of examples.

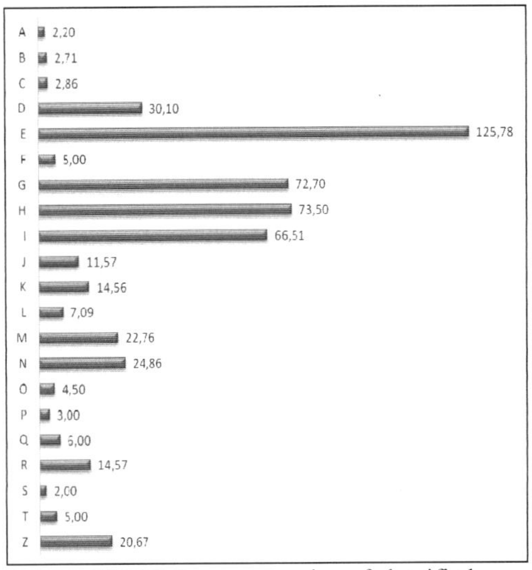

Fig. 11 Average number of classified diagnoses per cluster

Evaluation results (Table 3) shows high percentage of success in diagnoses recognition in PRs texts.

Table 3 Extraction sensitivity according to the IE performance measures

	Precision	Recall	F-measure
Diagnoses	97.3%	74.68%	84.5%

F-measure for leading diagnose in E-cluster and the diagnoses from the most common clusters (Fig. 9) is 98.76% for about 81.53% of the test set examples.

Obtained results are comparable with recent systems performing such task. For leading diagnoses we obtain better results, but still there are several difficulties like incorrect codes association due to:

- Latin terminology - for 345 cases;
- Abbreviations – for 538 cases
- Other – 1,202 cases describing mainly "status post" conditions, most of them is difficult to classify even manually.

For some diagnoses associated codes can be considered partially correct, because they the first three symbols of the ICD-10 code are assigned correctly, but the next tree symbols are either not specified, or associated to too general classes like *"... unspecified"*, *"... classified elsewhere"*, *"other disorders of..."* etc.

6 Conclusion and Further Work

This paper presents software modules for ongoing scientific project which supports the automatic extraction of diagnoses from PR texts

The implemented modules are strictly oriented to Bulgarian language.

Usage of SVM method for ICD-10 codes assignment to diagnoses, enhanced with some preprocessing techniques applied to input data, concerning usage of synonyms, hyponyms, negation processing, word normalization, Latin terminology and abbreviations processing and etc. shows better performance in certain context.

The plans for their further development and application are connected primarily to Bulgarian local context. For diagnoses recognition task we plan improvement of rules and extension of resource bank for Latin terminology and abbreviations for more precise code assignments.

Acknowledgments

The research tasks leading to these results have received funding from the EC's FP7 ICT under grant agreement n° 216130 PSIP (Patient Safety through Intelligent Procedures in Medication) as well as from the Bulgarian National Science Fund under grant agreement n° DO 02-292 EVTIMA (Efficient Search of Conceptual Patters with Application in Medical Informatics).

References

[1] International classification od Diseases, World Health Organization, http://www.who.int/classifications/icd/en/

[2] Demner-Fushman, D., W. Chapman and C. McDonald. What can natural language processing do for clinical decision support? *Journal of Biomedical Informatics, Volume 42, Issue 5, October 2009,* (2009), pp. 760-772.

[3] National Framework Contract between the National Health Insurance Fund, the Bulgarian Medical Association and the Bulgarian Dental Association, *Official State Gazette №106/30.12.2005, updates №68/22.08.2006 and №101/15.12.2006,Bulgaria*, http://dv.parliament.bg/

[4] 2007 International Challenge: Classifying Clinical Free Text Using Natural Language Processing, http://computationalmedicine.org/challenge/previous

[5] Pestian J, C. Brew, P. Matykiewicz, DJ Hovermale, N. Johnson, K. B. Cohen, and D. Wlodzislaw. A shared task involving multi-label classification of clinical free text. *In: ACL'07 workshop on biological, translational, and clinical language processing (BioNLP'07).* Prague, Czech Republic; (2007), pp. 36–40.

[6] Sotelsek-Margalef, A. and J. Villena-Román. MIDAS: An Information-Extraction Approach to Medical Text Classification (MIDAS: Un enfoque de extracción de información para la clasificación de texto médico), *Procesamiento del lenguaje Natural* n. 41, (2008), pp. 97-104.

[7] Hahn, U., M. Romacker and S. Schultz. Creating Knowledge Repositories from Biomedical Reports: The MEDSYNDIKATE Text Mining System, *In Pacific Symposium on Biocomputing,* vol. 7, (2002), pp. 338-349.

[8] Pakhomov, S., J. Buntrock and C. G. Chute. Automating the assignment of diagnosis codes to patient encounters, *Journal of American Medical Informatics Association*, 13, (2006), pp. 516-52.

[9] Coffman, A. and N. Wharton. Clinical Natural Language Processing: Auto-Assigning ICD- 9 Codes. Overview of the Computational Medicine Center's 2007 Medical Natural Language Processing Challenge. Available online at http://courses.ischool.berkeley.edu/i256/f09/Final%20Projects%20write-ups/coffman_wharton_project_final.pdf

[10] National Center of Health Information, http://www.nchi.government.bg/download.html

[11] 2011 ICD-10-CM Diagnosis Codes Index, http://www.icd10data.com/ and http://www.cdc.gov/nchs/icd/icd10cm.htm#10update

[12] T. Joachims. Text categorization with support vector machines: learning with many relevant features. In Claire N´edellec and C´eline Rouveirol, editors, *Proceedings of ECML-98, 10th European Conference on Machine Learning*, number 1398, Chemnitz, DE, Springer Verlag, Heidelberg, DE, (1998), pp. 137–142.

[13] Boytcheva, S. Assignment of ICD-10 Codes to Diagnoses in Hospital Patient Records in Bulgarian. In: Alfred, R., G. Angelova and H. Pfeiffer (Eds.). *Proc. of the Int. Workshop Extraction of Structured Information from Texts in the Biomedical Domain* (ESIT-BioMed 2010), ICCS-2010, Malaysia, (2010), pp. 56-66.

[14] Tcharaktchiev, D., G. Angelova, S. Boytcheva, Z. Angelov, and S. Zacharieva. Completion of Structured Patient Descriptions by Semantic Mining. In Koutkias V. et al. (Eds), *Patient Safety Informatics, Stud. Health Technol. Inform.* 2011 Vol. 166, IOS Press, (2011), pp. 260-269.

[15] Boytcheva, S., A. Strupchanska, E. Paskaleva, and D. Tcharaktchiev, Some Aspects of Negation Processing in Electronic Health Records. *In Proc. of Int. Workshop Language and Speech Infrastructure for Information Access in the Balkan Countries*, Bulgaria, (2005), pp. 1-8.

Evaluation Measures for Detection of Personal Health Information

Marina Sokolova
Faculty of Medicine,
University of Ottawa
and
Electronic Health Information Lab,
CHEO Research Institute
sokolova@uottawa.ca

Abstract

Texts containing personal health information reveal enough data for a third party to be able to identify an individual and his health condition. Detection of personal health information in electronic health records is an essential part of record de-identification. Performance evaluation in use today focuses on method's ability to identify whether a word reveals personal health information or not. In this study, we propose and show that the multi-label classification measures better serve the final goal of the record de-identification.

1 Introduction

Removing personal data in documents with sensitive contents aims to protect privacy of an individual from a third party and is called *de-identification* process. De-identification of electronic health records (EHR) became an important task of applied Health Informatics (Uzuner et al., 2007; Yeniterzi et al., 2010). Properly de-identified EHR, if revealed to a third party, will not identify the patient and his health conditions.

De-identification can be viewed as personal health information (PHI) detection, followed by alternation of the retrieved information (Danezis and Gurses, 2010). The first phase, PHI detection, uses Supervised Machine Learning, Natural Language Processing and Information Extraction techniques (Meystre et al., 2010). Name, date of birth, address, health insurance number are examples of PHI that should be detected:
[Name], [age], was admitted to the [Hospital] with chest pain and respiratory insufficiency. She appeared to have pneumonia ...

The present paper focuses on evaluation practices of PHI detection. Ordinary, the quality of PHI detection is measured in counts that record correctly and incorrectly recognized PHI words and word combinations. Table 1 presents a confusion matrix for binary classification; *tp* are true PHI, *fp* – false PHI, *fn* – false non-PHI, and *tn* – true non-PHI counts. *Accuracy*, *Precision*, *Recall*, *Fscore* are used to assess PHI detection (Yeniterzi et al., 2010). It is a common practice to evaluate detection as a binary classification of word categories (e.g., accuracy of name classification in a set of EHR).

Label \ Recognized	PHI	non-PHI
PHI	*tp*	*fn*
non-PHI	*fp*	*tn*

Table 1: A confusion matrix for binary PHI classification of words.

In this study, we argue that treating PHI detection as binary word classification does not fully meet the needs of the de-identification process. We instead formulate the PHI detection as a multi-label document classification, where performance is assessed through per-document multi-class classification. We propose that the multi-label document classification better serves the final goal of the EHR de-identification. We present a case study where *Exact Match Ratio*, *Labelling Fscore*, *Hamming Loss*, *One-error* are used to assess the PHI detection results.

2 Personal Health Information

A high demand in exchange and publishing of electronic health records promoted legislative actions of the patient privacy protection. In Ontario, Canada, the Personal Health Information Protection Act (PHIPA) protects the confidentiality of personal health information and the privacy of individuals with respect to that information, while facilitating the effective provision of health care [1].

[1] http://www.health.gov.on.ca/english/providers/legislation

Proceedings of the Workshop on Biomedical Natural Language Processing, pages 19–26,
Hissar, Bulgaria, 15 September 2011.

1. Names
2. All geographical subdivisions smaller than a State, including street address, city, county, precinct, zip code, and their equivalent geocodes, except for the initial three digits of a zip code
3. Dates (other than year) for dates directly related to an individual, including birth date, admission date, discharge date, date of death; and all ages over 89
4. Phone numbers
5. Fax numbers
6. Electronic mail addresses
7. Social Security numbers
8. Medical record numbers
9. Health plan beneficiary numbers
10. Account numbers
11. Certificate/license numbers
12. Vehicle identifiers and serial numbers, including license plate numbers;
13. Device identifiers and serial numbers;
14. Web Uniform Resource Locators (URLs)
15. Internet Protocol (IP) address numbers
16. Biometric identifiers, including finger, retinal and voice prints
17. Full face photographic images and any comparable images
18. Any other unique identifying number, characteristic, or code

Table 2: Health information protected by the Health Insurance Portability and Accountability Act (US).

Similar protection acts have been enabled: United States (US) has the Health Insurance Portability and Accountability Act [2], often known as HIPAA, European Union (EU) - Directive 95/46/EC, or, Data Protection Directive, although some details vary. Table 2 lists categories of personal health information which are protected by HIPAA.

Responsibility to protect patient's privacy promoted development of tools which de-identify electronic health records (EHR) (Morrison et al., 2009; Tu et al., 2010; Uzuner et al., 2007). First large-scale testing of de-identification tools showed that some of the protected categories do not appear in EHR (Uzuner et al., 2007). The absent categories included vehicle and device serial numbers, account numbers, internet protocol, URLs, and email. At the same time, references to health care providers (e.g., hospital, clinic) and professionals (e.g., doctors, nurses) frequently appeared and had been shown to reveal patient's health information. Table 3 summarizes the empirical evidence.

The de-identification systems usually benefit from the use of machine learning algorithms and text analysis methods (Meystre et al., 2010).

[2]http://www.hhs.gov/ocr/privacy/index.html

1. Age	4 Doctor	7 Location
2. Address	5 Hospital	8 Patient
3. Dates	6 ID	9 Phone

Table 3: PHI categories prevalent in EHR de-identification.

In practice, the PHI detection tools are usually trained and tested on same type of documents and/or documents originated from the same health care provider. They require a substantial amount of stored labeled training data and consume considerable time for its processing. We summarize relevant characteristics of the current PHI detection tools as follows:

- The goal of PHI detection is to detect personally identifiable information (e.g., name, address, age-identifying date).
- PHI detection applies to documents which are guaranteed to contain patient's health information (e.g., EHR).

- A common detection task is to identify whether a word bears PHI or not; for example, a phone number is PHI. Significant work was done to detect PHI indicators according to the HIPAA directives: detect and eliminate age-defining dates, postal codes, telephone numbers, social insurance numbers, etc.

3 Common Measures for PHI Detection

Currently, PHI detection methods are evaluated through their ability to correctly identify a PHI word category (e.g., *John* should be marked as a name) (Uzuner et al., 2007; Meystre et al., 2010; Morrison et al., 2009). This is done through assigning a word into two categories (i.e., binary classification), with more emphasis put on a correct labeling of PHI words.

Binary classification performance is the most general way of comparing the detection methods. It does not favour any particular application. The method's performance is assessed on all the input texts (e.g., correctly classified names in all the input discharge summaries). Introductions of new methods usually do not provide a detailed analysis of per-document detection results (Aberdeen et al., 2010; Gardner et al., 2010; Tu et al., 2010; Yeniterzi et al., 2010).

Focus on one class prevails in text classification, information extraction, natural language processing and bioinformatics. In those applications, the number of examples belonging to one class is often substantially lower than the overall number of examples. The same condition holds for the ratio of PHI words to all the words in electronic health records.

The PHI detection evaluation goes as follows: within a set of PHI categories there is a category of special interest (e.g., names). This category is designated as a *positive* class. The negative class is either *all other words* or *another PHI category*. The measures of choice calculated on the positive class are:

$$Precision = \frac{tp}{tp + fp} \tag{1}$$

$$Recall = \frac{tp}{tp + fn} \tag{2}$$

$$Fscore = \frac{(\beta^2 + 1)tp}{(\beta^2 + 1)tp + \beta^2 fn + fp} \tag{3}$$

All the three measures concentrate on the positive class (e.g., names):

Recall is a function of its correctly classified examples *tp* (e.g., *John* is classified as a name) and its misclassified examples *fn* (e.g., *John* is classified otherwise).

Precision is a function of *tp* and examples misclassified as positives (*fp*) (e.g., *Table* is classified as a name).

Fscore usually balances *Precision* and *Recall* with $\beta = 1$.

Accuracy does not distinguish between the number of correct labels of PHI and non-PHI classes. However, it approximates an over-all probability of correct classification (e.g., *John* classified as a name and *Table* is not classified as a name):

$$Accuracy = \frac{tp + tn}{tp + fn + fp + tn} \tag{4}$$

Further, we argue that performance measures other than per-class *Accuracy*, *Fscore*, *Precision*, *Recall* do apply to PHI detection and can be beneficial for EHR de-identification. Our argument focusses on the fact that the binary per-word classification leaves aside the quality of PHI detection in a document.

4 Paradox of high PHI detection and the PHI leakage risk

We claim that the currently reported PHI detection may not be sufficient to select methods which better prevent PHI leaks. We substantiate the claim by referring to the *re-identification risk*, i.e. the risk of identification of an individual from de-identified documents. The risk was actively studied for numerical data (El Emam et al., 2008).

Note that documents with undetected PHI can not have PHI altered, thus, cannot be completely de-identified. Thus, the re-identification risk depends on accuracy of PHI detection. At the same time, accuracy computed for word categories cannot solely account for EHR de-identification.

Let's have a set of records where every record contains three names: Patient, Location and Hospital. Suppose a de-identification method correctly detects 297 PHI indicators and misses 3. Consider two outcomes:

- if the three PHI indicators are missed in the same EHR, then that EHR poses a high risk of a unique patient identification;

- however, if the three indicators are missed in three separate EHR, then the re-identification risk is substantially lower.

To have a balanced picture of the document de-identification , we add another dimension, namely, distribution of missed PHI within the de-identified documents. Based on the missed PHI, we assign a de-identified document into the following re-identification groups:

high risk : a third party can identify an individual from the document content (e.g., Patient, Location, Hospital are not detected, hence, not de-identified);

medium risk : a third party needs one or two sources of additional information to identify an individual (e.g., Hospital is detected, but Patient and Location are not detected and not de-identified);

low risk : a third party needs several additional sources to identify an individual.

Section 5 presents a case study where a highly accurate PHI detection can still leak patient's health information through not properly de-identified EHR.

5 A case study

In this section, we illustrate the paradox of reporting binary classification results for PHI detection. Our scenario presents the situation in which three detection methods achieve similar error rates on every PHI category. Nevertheless, we further show that these methods are responsible for significantly different PHI leaks.

Let's have 500 EHR, where each document reports on one patient (e.g., referral letters, discharge summaries, lab reports). Detection methods A,B,C process the documents and obtain same scores in per-word classification. For each method, *Recall* ranges from 78.5% for Locations to 98.5% for Dates; the PHI indicators are missed as follows: Age – 1, Dates – 5, Doctor – 18, Hospital – 4, ID – 5, Location – 7, Patient – 5. [3]

Thus, we can conclude that the three methods are equally strong performers in PHI detection. But will this observation always be the case? We will now show that the superiority of one methods towards other methods largely depends on the applied evaluation measures. Let's assume that the three method errors were distributed considerably differently in per-document basis:

A in each document, A has missed no more than one PHI word; thus, there were 45 documents that had 1 missed PHI;

B if B misses a patient, then it misses a doctor, date, a location and ID in the same document; for other documents, B missed no more than one PHI word; thus, there were 5 documents with 5 different PHI missed and 20 documents with 1 missed PHI;

C always misses doctor names and another PHI in the same document; other PHI words were missed "one word per document"; thus, 18 documents missed 2 different PHI and 9 documents missed 1 PHI.

We can assume that the highest risk of patient identification comes from the 5 documents with original patient and doctor names, location indicators, numerical ID and dates. The lowest identification risk comes from the documents with one un-altered PHI example.

In terms of risk levels, A leaked 45 low risk documents. B leaked 25 documents with un-detected PHI, among those 5 high-risk documents and 20 low-risk documents. C leaked 27 documents, among those – 18 medium-risk documents and 9 low-risk documents. Table 4 presents the risks associated with every method.

Method	De-identified documents		
	High risk	Medium risk	Low risk
A	–	–	45
B	5	–	20
C	–	18	9

Table 4: Risks of the de-identified documents.

Binary classification results, *Recall* and error scores, do not differentiate between the three detection methods which contribute differently to re-identification risk prevention. Consequently, they may not lead to an appropriate selection of a detection method. To find a better selection approach, we recall that the PHI detection serves as the first step of the de-identification.

[3]These results and the number of words in each PHI category would be all the information necessary to compute the binary evaluation measures.

The detection goal, therefore, is to find so much of patient information in a document that its alteration will make the patient non-identifiable. This focusses us on two characteristics of a detection method:

1. the capability with respect to PHI word categories;

2. the ability to detect PHI categories within a given document.

We will now show that the two-dimensional PHI detection evaluation can be accommodated through the multi-labelling classification setting.

6 Multi-label classification

In multi-labelled classification, the document can be classified into several of l non-overlapping categories C_i (Sokolova and Lapalme, 2009). Examples include classification of functions of yeast genes (Mewes et al., 1997), identifying scenes from image data (Li et al., 2006), text-database alignment and word alignment in machine translation (Snyder and Barzilay, 2007), etc. In text mining of medical information, multi-label classification methods can be evaluated on OHSUMED, a collection of medical references (Hersh et al., 1994). When the learning task is document topic classification, multi-labelling is often referred as multi-topic classification (e.g., classification of clinical texts based on assigned multiple disease codes ICD-9-CM (Sasaki et al., 2007)).

The quality of multi-labelling classification is assessed through either partial or complete label matching (Kazawa et al., 2005); the latter is often referred to as exact matching. For an individual PHI category C_i, the assessment is defined by tp_i, fn_i, tn_i, fp_i. The following measures evaluate the performance on per-document:

- *Exact Match Ratio* (EMR) estimates the average per-document exact classification;

- *Labelling Fscore* (LF) estimates the average per-document classification with partial matches;

- *Hamming Loss* (HL) is the average per-document per-class total error;

- *One-error* (OE) estimates the proportion of documents with the mislabeled top label.

EMR, LF, HL count correct or incorrect label identification independently of their order or rank; OE counts incorrect labelling of the top ranked label.

In the formulae below, $L_i = L_i[1], \ldots, L_i[l]$ denotes a set of class labels for x_i, $L_i[j] = 1$ if C_j is present among the labels and 0, otherwise; L_i^{class} are labels given by a method, L_i^{data} are the document labels; L_i^t is the top ranked label, I is the indicator function.

$$EMR = \frac{1}{n}\sum_{i=1}^{n} I(L_i^{class} = L_i^{data}) \quad (5)$$

$$LF = \frac{2}{n}\sum_{i=1}^{n}\sum_{j=1}^{l} \frac{L_i^{class}[j]L_i^{data}[j]}{(L_i^{class}[j] + L_i^{data}[j])} \quad (6)$$

$$HL = \frac{1}{nl}\sum_{i=1}^{n}\sum_{j=1}^{l} I(L_i^{class}[j] \neq L_i^{data}[j]) \quad (7)$$

$$OE = \frac{1}{n}\sum_{i=1}^{n} I(L_i^t \neq L_i^t) \quad (8)$$

For EMR and LF, higher values show a better match between input and output labels. For HL and OE, the reverse is true.

7 PHI detection as a multi-label classification

We formulate PHI detection as a multi-label classification problem, where labels represent PHI categories. In this case, a document is assigned a label if the corresponding PHI category is found in the document contents.

Let's consider three labels: a name, a location, and the "other PHI" (e.g.,a doctor, ID, age). Knowing all the three pieces of information allows for identification of an individual, i.e., represents a high re-identification risk. We assume that the input labels are set to 1 as EHR contains the patient information in all the three categories.

If a method properly detects the three categories, then the output EHR should not contain that information. Hence, all the three output labels should be 0. If a PHI word is not detected and the information leaks out, then the EHR output label for the corresponding category is 1. This implies that a poorer PHI detection is signaled by a bigger match between the input and output labels. A smaller match between the labels signals otherwise.

In terms of the measures introduced in Section 6, we interpret their values as follows:

EMR is 0 if there is no EHR with PHI missed in all the three categories; if $EMR > 0$, then there is at least one EHR with undetected PHI in all the three categories;

LF is higher when there are more EHR with several undetected PHI;

HL is lower when there are more EHR with undetected PHI;

OE is lower when more EHR contain the top PHI undetected.

We apply the multi-label measures to evaluate A,B,C performance given in Section 5. To find the top ranked label, we note that the geographic information has the biggest impact on person re-identification (Herzog et al., (2007); El Emam et al., 2008). Thus, the location is designated as the top PHI category; its detection results are used to compute OE. As an intermediate step, Table 5 reports the exact and partial label matches for the three methods. The measure values are reported in Table 6.

Method	$I(L_i^{class} = L_i^{data})$	$\sum_{j=1}^{l} \frac{L_i^{class}[j] L_i^{data}[j]}{(L_i^{class}[j] + L_i^{data}[j])}$	$\sum_{j=1}^{l} I(L_i^{class}[j] \neq L_i^{data}[j])$	$I(L_i^t \neq L_i^t)$
A	–	11.25	1455	493
B	5	7.50	1465	493
C	–	9.45	1455	493

Table 5: Counts of exact and partial label matches for A,B,C.

Method	Multi-label measures			
	EMR	LF	HL	OE
A	0.00	0.045	0.97	0.986
B	0.01	0.030	0.98	0.986
C	0.00	0.038	0.97	0.986

Table 6: Multi-label evaluation of methods A,B,C

EMR, the win-or-loose measure, shows that B outputs EHR with a high re-identification risk. OE shows that A,B,C output the same volume of documents in which the top ranked PHI was undetected. LF is the most discriminative measure among those applied: it marks B as the most unsafe detector, C – as a distant second, and A - as the safest detection method .

This empirical comparison shows that the A,B,C performance is not equivalent for the PHI detection, although the binary classification measures led us to believe otherwise in Section 5. In fact, the method performance can be considered significantly different if the re-identification risk is taken into account. We have shown that the multi-label classification measures account for that difference: LF differentiated between the three methods, EMR, HL differentiated between B and A, C, although they marked the performance of A,C as equivalent; OE illustrated that the three methods equally missed the location information.

In terms of the re-identification risk, EMR singles out methods with potentially higher re-identification risk, LF separates the high, medium and low risk methods; OE concentrates on the most important category; evaluation by HL is more subtle.

Efficiency of multi-labelled classification has been discussed by Kazawa et al (2005), Fujino et al (2008), Mencia et al (2010). They showed that classification costs depend on the prior knowledge about the labels (e.g., established correspondence between training and test labels) and are proportional to the number of label categories per example. We leave the analysis of efficiency of multi-labelled PHI detection for future work.

8 Related Work

Several PHI detection tools were developed and deployed to process EHR data. These tools focus on retrieval of patient's personally identifiable information, such as patient's name, address, the name and address of the health care provider or insurer. The tools de-identify clinical discharge summaries (Uzuner et al., 2008), nurse notes (Neamatullah et al., 2008), pathology reports (Beckwith et al., 2006). So far, the published work on PHI de-identification reports results in terms of binary classification. *Precision*, *Recall*, *Fscore*, *Accuracy* are reported for

PHI categories (e.g., name, address) but not for per-document performance.

A common presentation of a PHI detection method would include the use of dictionaries of local personal and geographic names. For example, in (Neamatullah et al., 2008), the authors built a system to detect PHI in nurse notes. Manual de-identification of the notes is highly accurate: the averaged manual *Precision* = 98.0%. To improve the automated de-identification, the authors use customized dictionaries of local person, geographic and health care provider names. Without the localized dictionaries, the tool's overall *Precision* is 72.5%. When the dictionaries are used, the tool's overall *Precision* is 74.9%. The tool's performance substantially varies on identification of individual categories. For person names, the use of the customized dictionaries is adverse: *Precision* = 73.1% without the dictionaries and 72.5% – with them. Location detection, in contrast, considerably improves with the use of the local dictionaries: *Precision* increases from 84.0% to 92.2% when the local information is available, *Recall* – from 37.0% to 97.0%. The use of customized dictionaries of local names, health care providers, acronyms and "do not remove" medical terms was shown to improve the PHI detection on heterogenous EHR, gathered from several regional clinics (Tu et al., 2010). The reported *Fscore* increases from 77%, without the use of customized dictionaries, to 90%, when the dictionaries are used.

Testing detection methods on altered versions of same documents is another common trend in evaluation. EHR de-identification systems are commonly trained on re-synthesized records, i.e. records where real identifiers are substituted by synthetic ones. The re-synthesis effects on personal information detection were studied in (Yeniterzi et al., 2010); the researchers used the de-identification system first introduced by (Aberdeen et al., 2010). The system's *Fscore* declined from 98.0%, when tested on re-synthesized records, to 72.8% when tested on original records. When trained on records with original PHI, the system's performance fluctuated less: *Fscore* was 96.0%, when tested on original records, and 86.2%, when tested on re-synthesized records.

The reported accuracy, however high, does not provide enough data for a thorough understanding of the PHI de-identification. We suggest to incorporate the re-identification risk in the reported results. This can be done, for example, through per-document performance evaluation.

Relations between disclosed parts of personal information are studied for social networks. In (Al-Faresi et al, 2010), the authors apply Bayesian networks to model the risk of re-identification from email and a forum post. In (Domingo-Ferrer, 2009; Domingo-Ferrer and Saygin, 2009), the authors discuss privacy risk scores where private data categories are assigned sensitivity weights. The authors do not report how the weights should be calculated or what private categories are suggested. De-identification and re-identification processes are also left out of the study scope.

9 Conclusions and Future Work

Prevention of patient PHI leaks into a public domain has traditionally been an obligation of those responsible for the safeguarding of the information. Such obligation, reinforced by legislative requirements, prompted development of PHI de-identification methods and tools. Machine Learning, Natural Language Processing and Information Extraction techniques became essential parts of the de-identification process.

In this study, we have proposed a new approach to the evaluation of PHI detection, the first part of the de-identification. Our approach focuses on the method's ability to detect the PHI information within a document. We have argued that this not yet been done in studies of PHI detection. We proposed performance measures which originate in multi-label classification: *Exact Match Ratio, Labelling F-score, Hamming Loss, One-error.* Our case study of electronic health record de-identification has presented an application which benefits from the use of these measures. We also have presented PHI detection as the multi-label classification problem.

Our future work will follow several interconnected avenues: incorporate PHI alteration in free-form texts, find new characteristics of the methods that must be evaluated, consider new measures of method performance, and search for PHI detection applications other than EHR de-identification.

Acknowledgments

This work has been supported by NSERC and CIHR operating research grants.

References

Aberdeen, J. Bayer, S., Yeniterzi, R., Wellner, B., Clark, C., Hanauer, D. , Malin, B., Hirschman, L. The MITRE Identification Scrubber Toolkit: Design, training, and assessment. *International Journal of Medical Informatics*, 2010. 79(849–859)

Al-Faresi, A., A. Alazzawe, A. Alazzawe, and Duminda Wijesekera. Risk Analysis Framework & Architecture for DLP Systems, *Proceedings of IMPPCD-2010*, p.p. 35–43, 2010. http://www.ehealthinformation.ca/documents/IMPPCD-2010.pdf

Beckwith, B., R. Mahaadevan, U. Balis, and F. Kuo, Development and evaluation of an open source software tool for de-identification of pathology reports. *BMC Medical Informatics and Decision Making*, 2006; 6:12;

Danezis, G. and S. Gurses. A critical review of 10 years of Privacy Technology. In *the Proceedings of Surveillance Cultures: A Global Surveillance Society?*, 2010.

Domingo-Ferrer, J. The Functionality-Security-Privacy Game. *Proceedings of MDAI*, 2009, 92–101.

Domingo-Ferrer, J. and Y. Saygin. Recent progress in database privacy. *Data and Knowledge Engineering*, **68** (11), 1157–1159, 2009.

El Emam,K., A. Brown, P. Abdel Malik. Evaluating Predictors of Geographic Area Population Size Cut-offs to Manage Re-identification Risk. *Journal of American Medical Informatics Association*, **16**(2), 256–266, 2008.

Fujino, A., H. Isozaki, J. Suzuki. Multi-label Text Categorization with Model Combination Based on F1-score Maximization. *Proceedings of IJCNLP*, 2008, 823–828.

Gardner, J., L. Xiong, F. Wang, A. Post, J. Saltz, T. Grandison. An Evaluation of Feature Sets and Sampling Techniques for De-identification of Medical Records. *Proceedings of IHI*, 2010, 183–190.

Hersh, W., C. Buckley, T. Leone, and D. Hickam, 1997. OHSUMED: An interactive retrieval evaluation and new large test collection for research. In *Proceedings of the 17th Annual International ACM SIGIR Conference on Research and Development in Information Retrieval (SIGIR-97)*, pp. 192-201

Herzog, T., F. Scheuren, and W. Winkler, *Data Quality and Record Linkage Techniques*. 2007: Springer

Kazawa, H., Izumitani, T., Taira, H., and E. Maeda, 2005. Maximal margin labeling for multi-topic text categorization. In *Advances in Neural Information Processing Systems*, Vol. 17. pp. 649–656.

Li, T., Zhang, C., and S. Zhu, 2006. Empirical studies on multi-label classification. In *Proceedings of the 18th IEEE International Conference on Tools with Artificial Intelligence*, pp. 86–92.

Mencia, E., S.-H. Park, J. Furnkranz. Efficient voting-predictionforpairwisemultilabelclassification. *Neurocomputing*, **73**, p.p. 1164–1176, Elsevier, 2010.

Mewes, H.-W., K. Albermann, K. Heumann, S. Lieb, and F. Pfeiffer, 1997. MIPS: a database for protein sequences, homology data and yeast genome information. *Nucleic Acids Research*, 25(1), p.p. 28-30.

Meystre, S., F. Friedlin, B. South, S. Shen, and M. Samore. Automatic de-identification of textual documents in the electronic health record: a review of recent research. *BMC Med Res Methodol*. 2010; 10: 70

Morrison, F., et al., Repurposing the Clinical Record: Can an Existing Natural Language Processing System De-identify Clinical Notes? *Journal of American Medical Information Association*, 2009. 16(1):37-39;

Neamatullah, I., et al., Automated De-Identification of Free-Text Medical Records. *BMC Medical Informatics and Decision Making*, 2008; **8**(32);

Sasaki, Y., Rea, B., and S. Ananiadou, 2007. Multi-topic Aspects in Clinical Text Classification. In *Proceedings of the 2007 IEEE international Conference on Bioinformatics and Biomedicine*, IEEE Computer Society, pp. 62–70.

Sokolova, M. and G. Lapalme. A systematic analysis of performance measures for classification tasks, *Information Processing and Management*, **45** (4), 427–437, 2009.

Snyder, B., and R. Barzilay, 2007. Database-text Alignment via Structured Multilabel Classification, In *Proc. of the International Joint Conference on Artificial Intelligence (IJCAI-2007)*, pp. 1713–1718.

Tu, K., J. Klein-Geltink, T. Mitiku, C. Mihai and J. Martin. De-identification of primary care electronic medical records free-text data in Ontario, Canada.; *BMC Medical Informatics and Decision Making*, 2010; 10:35;

Uzuner, O., Y. Luo, and P. Szolovits, Evaluating the state-of-the-art in automatic de-indentification. Journal of the American Medical Informatics Association, 2007; 14: 550–563.

Uzuner, O., et al., A de-identifier for medical discharge summaries. *Journal of Artificial Intelligence in Medicine*, 2008; 42:13–35;

Yeniterzi, R., Aberdeen, J., Bayer, S., Wellner, B., Hirschman, L., Malin, B. Effects of personal identifier resynthesis on clinical text de-identification. *Journal of American Medical Information Association*, 2010. 17: 159–168

Building a Named Entity Recognizer in Three Days: Application to Disease Name Recognition in Bulgarian Epicrises

Georgi D. Georgiev
Ontotext AD
IT Center Office Express, 3rd floor, 135
Tsarigradsko Shosse, Sofia 1784, Bulgaria
georgi.georgiev@ontotext.com

Valentin Zhikov
Ontotext AD
IT Center Office Express, 3rd floor, 135
Tsarigradsko Shosse, Sofia 1784, Bulgaria
valentin.zhikov@ontotext.com

Borislav Popov
Ontotext AD
IT Center Office Express, 3rd floor, 135
Tsarigradsko Shosse, Sofia 1784, Bulgaria
borislav.popov@ontotext.com

Preslav Nakov
National University of Singapore
Department of Computer Science
13 Computing Drive, Singapore 117417
nakov@comp.nus.edu.sg

Abstract

We describe experiments with building a recognizer for disease names in Bulgarian clinical epicrises, where both the language and the domain are different from those in mainstream research, which has focused on PubMed articles in English. We show that using a general framework such as GATE and an appropriate pragmatic methodology can yield significant speed up of the manual annotation: we achieve F1=0.81 in just three days. This is the first step towards our ultimate goal: named entity normalization with respect to ICD-10.

1 Introduction

The problems of named entity recognition and normalization are central to biomedical text processing: as part of the typical preprocessing pipeline, they are key for any deep text analysis.

The goal is to identify all mentions of named entities of a particular type, e.g., genes, proteins, diseases, drugs, and to propose a canonical name, or a unique identifier, for each mention. Solving this problem is important for many applications, e.g., enriching databases such as the *Protein and Interaction Knowledge Base*[1] (PIKB), part of *LinkedLifeData*, compiling gene-disease-drug search indexes for large document collections in *KIM*[2] (e.g., using the BioMedicalTagger[3]) and *MEDIE*[4], or building a search engine that can retrieve the effects of a drug on various diseases in research papers and patents.

Being so central to biomedical text processing, the problems of named entity recognition and normalization have received a lot of research attention, e.g., there have been several related competitions at BioNLP[5] and BioCreAtIvE[6]. Moreover, high-quality manually annotated biomedical text corpora such as GENIA[7] have been created, which have enabled the development of a number of biomedical text processing tools that need such kind of data for training.

Unfortunately, mainstream research has so far focused almost exclusively on English and on biomedical abstracts and full-text articles in PubMed. Thus, biomedical named entity recognition (NER) for languages other than English or for other types of biomedical texts faces the problem of the lack of manual text annotations and biomedical resources in general, which are needed for machine learning. While manually annotating some data is always a good idea, e.g., for analysis, parameter tuning and evaluation, it is hardly practical for more than just a few documents. It is thus important to make smart use of any existing resources and to facilitate the process of manual annotation as much as possible so that good results can be achieved quickly and with very little efforts. The best approach depends on the particular task as well as on the kinds of texts and resources that are available, and there is hardly a universal solution. Still, there are probably lessons to be learned from particular examples of efforts focusing on achieving good performance for NER in new languages and domains in a short period of time.

[1] http://www.linkedlifedata.com
[2] http://www.ontotext.com/kim
[3] http://www.ontotext.com/life-sciences/semantic-biomedical-tagger
[4] http://www-tsujii.is.s.u-tokyo.ac.jp/medie/
[5] http://sites.google.com/site/bionlpst/
[6] http://biocreative.sourceforge.net/
[7] http://www-tsujii.is.s.u-tokyo.ac.jp/GENIA/home/wiki.cgi

Proceedings of the Workshop on Biomedical Natural Language Processing, pages 27–34,
Hissar, Bulgaria, 15 September 2011.

Below we describe our experience with building a recognizer for disease names in Bulgarian clinical epicrisis, where both the language and the domain are different from those in mainstream research. We demonstrate good performance with little efforts and in a very short period of time. We further show how we can save about 57% of the efforts needed for manual annotation of instances of named entities in text. Finally, we discuss how this work can be extended to the task of named entity normalization.

The remainder of the paper is organized as follows: Section 2 offers an overview of related work, Section 3 summarizes our methodology, Section 4 describes our experiments and presents the results, Section 5 goes into deeper analysis, Section 6 describes some potential applications, and Section 7 concludes with possible directions for future work.

2 Related Work

There have been several research efforts focused on making manual annotations over a text corpus or building a system for named entity recognition and information extraction in a short period of time and with limited resources.

Ganchev *et al.* (2007) proposed a semi-automated approach to named entity annotation where first an *n*-best MIRA-based named entity recognizer is trained on the initial training set and then tuned for high recall by manipulating the MIRA loss function. Then, its output is checked by a human annotator, who makes yes/no decisions for each proposed entity. Their experiments show that this can speed up manual annotation by about 58% without loss in quality of annotation. We achieve a similar reduction of 57% in the required annotation time using a structured perceptron for named entity tagging; however, we start with no annotated data at all.

Settles (2011) described the DUALIST system for semi-supervised annotation based on active learning. The system solicits and learns from labels on both features (e.g., words) and instances (e.g., entities). It has been evaluated on a number of annotation and classification tasks; on named entity recognition, it achieved 0.80 precision (unknown recall and F1), which is a bit lower than our 0.86-0.87 precision. Moreover, being based on active learning, DUALIST needs access to a large number of unlabeled documents from which to choose examples for annotation; such documents are not available in our case.

Freedman *et al.* (2011) presented a bootstrapping system for what they call *extreme extraction*, where they start with an ontology defining the target concepts and relations they will need to extract and a limited number of training data. They achieve human-level accuracy in a week; this includes five hours of manual rule writing. In fact, our case is arguably more extreme since we start with no annotated data at all.

We should also mention the early work done as part of the Surprise Language Exercises held in 2003, where sixteen teams tried to develop language technologies for two previously unanticipated languages, Cebuano and Hindi, in just ten and twenty-nine days, respectively. This work is described in two special issues of the *ACM Transactions on Asian Language Processing* journal (Oard, 2003).

Finally, we should mention the 2007 Computational Medicine Challenge[8], which focused on analyzing clinical epicrises but for English. However, it asked for assigning ICD-9 codes at the document level, while we want to find *instances* of *IDC-10* disease mentions in text. Moreover, the challenge provided a lot of manually curated data, and thus there was no need to annotate additional data (Crammer et al., 2007).

3 Method

We started with a small number of Bulgarian epicrises and a list of diseases from an ontology.

First, we analyzed and manually annotated a small number of documents to acquaint ourself with the data and the task and to produce datasets for development and evaluation.

Next, we automatically induced contextual rules for finding additional names of diseases. We applied these rules on the development set, we inspected their output, and we incrementally restricted them in several iterations. Once we were satisfied with the precision, we applied the rules to new texts, and we collected their predictions to build a gazetteer of likely disease names.

Next, we applied the gazetteer to some unannotated documents, and we inspected and corrected the matches in context, thus ending up with more annotated documents for training. We then trained a sequence-based named entity recognizer on all training documents we had so far.

Finally, we augmented that sequence recognizer with the predictions of the gazetteer as features, thus achieving 0.81 F1 score, which we found sufficient, and we stopped there.

[8] http://computationalmedicine.org/challenge/previous

4 Experiments and Evaluation

4.1 Initial Datasets

We started with a collection of 100 Bulgarian documents describing clinical epicrises, which we analyzed manually. Based on this analysis, we developed annotation guidelines, which we used to annotate 20 of these documents with the names of diseases from the ICD-10[9] (International Classification of Diseases) ontology. There were a total of 441 disease names mentioned in these 20 documents.

4.2 Contextual Rule Induction

We selected 10 of the annotated documents for development, and used the remaining 10 documents for testing.

We automatically learned contextual rules from the development set, which we then used to find named entities in the test set.

The rules memorize three tokens to the left and to the right around a potential disease name instance. Here is an example (`???` marks the target instance):

```
diagnosis : ??? . Polyneuropathia Diabetica
```

We do not allow the context words to cross sentence boundaries, and thus the inferred rules can have access to a context of less than three words on either or both sides of `???`, as in the following example:

```
the therapy of ??? .
```

Here are some inferred rules extracted from the original text in Bulgarian:

```
ДИАГНОЗА : ??? . ПОЛИНЕВРОПАТИЯ ДИАБЕТИКА
. ??? . ХИПОТИРЕОИДИЗМУС АУТОИМУНЕС
. ??? . ХИПЕРТОНИЯ АРТЕРИАЛИС
. II . ??? . ПИЕЛОНЕФРИТИС ХРОНИКА
. ??? .
във връзка със ??? , диагностично уточняване
степента на ??? и лечение .
Минали заболявания : ??? от 20 години
стабилна стенокардия , ??? от 25 години
млада възраст , ??? и еритема нодозум
5 години , ??? - хипотиреоидна фаза
ЕМГ данни за ??? . Намалена скорост
за лечение на ??? . Проведе се
в терапията на ??? .
```

[9] http://www.who.int/classifications/icd/en/

Rules with a balanced context of three words on either side proved to be restrictive and very reliable. We indexed the annotated documents in GATE (Cunningham, 2000) so that we could perform fast search for annotations and context words on either side of a disease mention.

The rules were implemented in JAPE format[10]:

```
Rule: One
Priority: 100(
      ({Token.string   == "ДИАГНОЗА"})
      ({Token.string   == ":"})
      (({Token})+):bind
      ({Token.string   == "."})
-->
      :bind.PreDisease = {rule = "One"}
```

The preceding rule states that if the word "diagnosis" is followed by ":", all following tokens up to and not including "." should be considered part of a disease name.

Figure 1 shows the user interface for searching and visualization of the results in the context of three tokens to the left/right of the candidate disease names. Fast searching allows us to find that two tokens on the right hand side and only one on the left hand side could yield better results.

4.3 Inducing a Disease Gazetteer

We executed the rules on the 80 unannotated documents and we created a gazetteer based on the recognized disease names. The resulting gazetteer was evaluated based on its ability to find correct disease mentions in text, on the development set and on the test set. The results are shown in Table 1.

	R	P	F1
Dev set	0.61	0.30	0.40
Test set	0.61	0.49	0.54

Table 1: Evaluation of the gazetteer that was induced using the context rules.

As Table 1 shows, the low precision is a more important problem for the induced gazetteer than low recall. We thus focused on improving precision by adding rules that could filter out some bad extracted candidates.

We first experimented with length-based filters, removing all candidates whose length is less than n symbols; we tried n = 5, 6, 7, 8, 9, 10. The results are shown in Table 2. We can see that using length filters significantly increases precision without negatively affecting recall: precision jumps from 0.3 to 0.65, which is higher than recall. As a result, F1 raises from 0.4 to 0.61.

[10] http://gate.ac.uk/sale/tao/index.html#x1-2030008

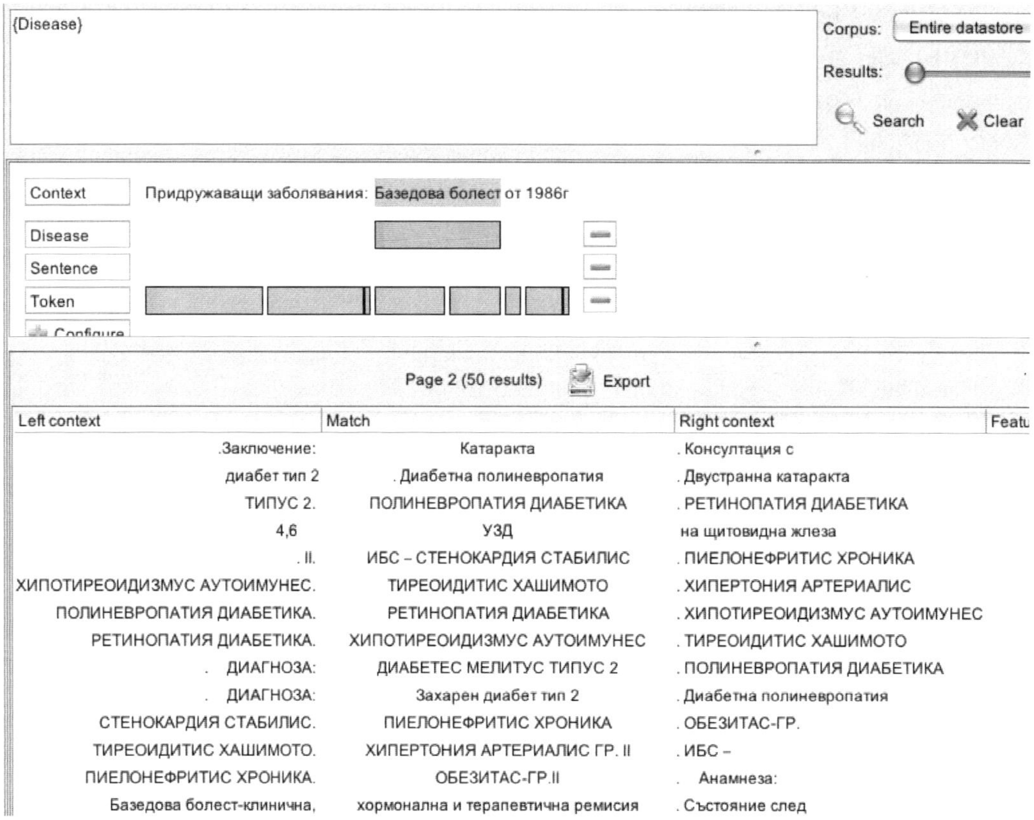

Figure 1: The user interface used to manually annotate instances.

Filter	R	P	F1
Length < 5	0.61	0.47	0.53
Length < 6	0.61	0.54	0.57
Length < 7	0.60	0.60	0.60
Length < [8-10]	0.58	0.65	0.61
+ "^[0-9] .*"	0.56	0.72	0.62
+ specific words	0.56	0.87	0.68

Table 2: Evaluation (on the *development dataset*) of the gazetteer that was induced from context rules *and additional filtering rules*.

Next, we looked closer at the development dataset, and we found that many candidate disease names that started with numbers were in fact false positives. Thus, we tried adding a regular expression like "^[0-9] .*" as an additional filter to the length filter of 8. This improved precision from 0.58 to 0.72, but recall dropped from 0.60 to 0.56, and thus, F1 increased only slightly: from 0.61 to 0.62.

We further tried removing candidates containing certain words like *menopause* and *unencumbered*, which our manual analysis has found to give rise to many false positives. Doing so led to another increase in precision to 0.87 on the development set while keeping the recall intact at 0.56. The F1 score for the gazetteer after these rules were applied increased from 0.62 to 0.68 (R=0.56, P=0.87) on the development set. The same F1 of 0.68 was also achieved on the test set, but there recall and precision were more balanced: R=0.64, P=0.73.

These results are arguably not strong enough for a system that recognizes disease mentions in clinical epicrisis in a fully automated fashion. Still, the resulting gazetteer could potentially be useful for a number of tasks, e.g., for making more manual annotations faster. It could also help robustness since it captures most of the rules that were defined in the annotation guidelines created during the first run of the manual annotation: the 20 documents that we used for development and testing.

Thus, we decided to use the gazetteer to help the annotation of 10 more documents. The annotation required about 1.6 minutes per document for the one annotator and 3.6 minutes for another one, or 2.6 minutes on average. As a comparison, in the first annotation run, the average time for annotating a document was about 6 minutes for both annotators. This is a 57% reduction in the time needed to annotate a new document.

There was also an improvement in the quality of the manual annotation process. The average agreement between the gazetteer and the manual annotators was F1=0.84 (R=0.88, P=0.80) for the first annotator and 0.77 (R=0.67, P=0.91) for the second one. The inter-annotator agreement was F1=0.78 (R=0.88, P=0.70). This is comparable to the inter-annotator agreement between the two annotator on the first 10 documents where F1 was 0.80 (R=0.96, P=0.69) and to the second 10 where F1 was 0.77 (R=0.91, P=0.67).

4.4 Training a Structured Perceptron for Disease Name Recognition in Text

The resulting 30 documents (20 for training and 10 for testing) were used to train a structured perceptron (Freund and Shapire, 1999). This learning algorithm was selected for simplicity and because of its fast online training. We used a standard set of features that has been initially proposed by McDonald & Pereira (1996), and then successfully adapted to Bulgarian by Georgiev *et al.* (2009); shown in Table 3.

Using this feature set, the perceptron achieved an F1 of 0.69, which is only slightly better than the 0.68 F1 of the rules/gazetteer approach.

In order to improve the performance, we annotated 10 more documents by first applying the gazetteer and then doing manual annotations to create System 1, which was trained on 30 documents and tested on 10. We further built System 2, which was trained as System 1, but also used matches with the gazetteer as features.

The results for System 1 in Table 4 show that adding more training data (i.e., 30 instead of 20 documents) yields only minor improvement in F1: from 0.69 to 0.71. However, also using features from the gazetteer in System 2 causes F1 to jump to 0.81. As we can see, this is due to a huge improvement in recall, which goes from 0.59 to 0.76, while precision remains stable.

We can conclude that the gazetteer turned out to be an important information source, probably because it had analyzed more text (90 documents; all but the testing 10 ones), and thus it could help recall a lot.

Predicate	Regular Expression
Initial capital	[А-Я].*
Capital, then any	[А-Я].
Initial capitals, alpha	[А-Я][а-я]*
All capitals	[А-Я]+
All lowercase	[а-я]+
Capitals mix	[А-Яа-я]+
Contains a digit	.*[0-9].*
Single digit	[0-9]
Double digit	[0-9][0-9]
Natural number	[0-9]+
Real number	[-0-9]+[\.,]?[0-9]+
Alpha-numeric	[А-Яа-я0-9]+
Roman	[ivxdlcm]+|[IVXDLCM]+
Contains dash	.*-.*
Initial dash	-.*
Ends with dash	.*-
Punctuation	[,\.;:\?!-+"]
Multidots	\.\.+
Ends with dot	.*\.
Acronym	[А-Я]+
Lonely initial	[А-Я]\.
Single character	[А-Яа-я]
Quote	["’]

Table 3: The orthographic predicates used by the structured perceptron named entity recognizer. The observation list for each token includes a predicate for each regular expression that matches it.

System	R	P	F1
System 1	0.59	0.87	0.71
System 2	0.76	0.86	0.81

Table 4: Evaluation of the structured perceptron. System 1 is trained on 30 documents and tested on 10. System 2 is trained like System 1, but it also uses features based on matches with the gazetteer.

5 Discussion

The experiments above have shown that manually annotating data and building a system for named entity recognition in clinical epicrises written in Bulgarian is hard for a number of reasons, including but not limited to the following:

(i) limited and chaotic general purpose text analysis resources for Bulgarian,
(ii) our lack of experience with such texts,
(iii) specificity of the domain language,
(iv) specificity of the terminology,
(v) specificity of the document structure,
(vi) issues with extracting the text from the Microsoft Word format the epicrises were stored in.

We should note that extracting disease names from epicrises would hardly have been much simpler for English, despite the existence of many biomedical corpora and tools for that language. The main problem here is the domain shift: the existing tools and resources for English are targeting almost exclusively journal papers, whose format, structure and vocabulary differs substantially from those of clinical epicrises.

On the positive side, we have shown that even though the task looks complicated, it could be solved with usable F1 in just three days.

This speed of building our system would have hardly been possible without our extensive use of the GATE framework for natural language engineering, which has saved us a lot of time and efforts. Among its features that have helped us the most were *(i)* its ability to extract text from Microsoft Word documents, *(ii)* its default Unicode tokenizer and *(iii)* its sentence splitter based on simple regular expressions, which we were able to adopt very quickly, thus overcoming the lack of general purpose text analysis tools and resources for our kind of biomedical text.

We were further able to speed up the process of manual annotation by focusing on rules based on words/tokens rather than on part of speech or lemmata (for which we did not have ready tools that could handle the domain well). This was possible because of the particular structure of the documents and the specific language use.

For example, clinicians tend to express the diagnosis at the beginning of the epicrisis, typically, in a paragraph that starts with the pattern "Diagnosis:" (or "Диагноза:" in Bulgarian). Here is an example:

ДИАГНОЗА: **Захарен диабет-тип 2. Артериална хипертония. Дислипидемия.**

The diagnosis is followed by few paragraphs explaining why and how the patient was examined, which is further followed by additional information about how the presence of the disease was tested.

Of course, a diseases can be extracted from other parts of the document, e.g., such that provide information about the examination of the patient by another specialist. For example, the structured paragraph below contains a list of diseases that have been suggested after a consultation with a neurologist:

Консултация с невролог: **Начален полиневропатен синдром.** Терапия: контрол на кръвната захар.

We should note that disease names can be mentioned not only in the diagnosis-related section(s) of an epicrisis, but can occur pretty much anywhere in the document. While catching all instances is generally hard, we were able to do it with the high F1 of 0.81 to a great extent because of the gazetteer. This is because most of the disease names mentioned outside of the diagnoses are likely to repeat those that have been already listed in the diagnosis section. Thus, once the gazetteer has been populated with the somewhat easy-to-extract disease names from the diagnosis-related sections, it can help find further instances of those disease names in contexts that are generally much more ambiguous. We have seen this effect above when comparing System 1 and System 2 in Table 4.

6 Potential Applications

As we have seen above, System 2 recognizes disease mentions in text with an F1 of 0.81, which is quite high and is arguably already usable for a number of practical applications. Still, generally, named entity recognition is just the first step in biomedical text analysis; we might also need normalization, which would allow us to get to the canonical names of the diseases mentioned in a particular epicrisis, thus enabling more sophisticated practical applications. For example, if a disease recognizer is coupled with a recognizer of dates and symptoms, we would be able to monitor disease progression and manifestation over time.

Normalizing disease names to an identifier or a canonical name in an ontology, would also allow linking a particular clinical epicrisis to a whole web of linked data. One such example would be *LinkedLifeData*, which is a platform that integrates biomedical information for diseases, symptoms, proteins, genes, drug action information and clinical trials. Linking between an epicrisis and LinkedLifeData might facilitate knowledge acquisition and enrichment and could enable sophisticated queries and rich semantic search over a collection of epicrises.

Thus, in order to enable such semantic annotations, we need not only the offsets and type of each disease mention in a given epicrisis but also a mapping of the mention to a unique identifier. An obvious candidate in our case is the Bulgarian version of the ICD-10 ontology, which provides both unique disease name identifiers and canonical forms that can be used for disease name normalization.

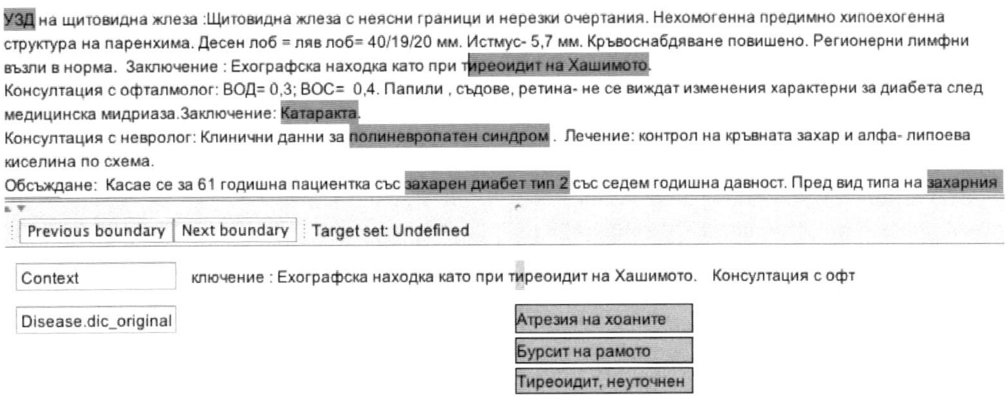

Figure 2: The GATE user interface for choosing the correct canonical names for the disease instances in text.

Motivated by the practical importance of the task, we did some preliminary assessment of the feasibility of the idea of mapping disease name mentions to identifiers in the ICD-10 ontology. Unfortunately, we found that this was not as simple as we thought initially since the disease names used in ICD-10 strongly disagreed with the names used in our clinical epicrises.

One reason for this is the tendency of Bulgarian clinicians to describe their diagnoses in Latin. Unfortunately, the Bulgarian ICD-10 does not include disease names in Latin. Moreover, there were many abbreviations, both for Latin and Bulgarian. Thus, for this task, it is important to collect Latin medical terminology from other sources as well as synonyms of diseases for Bulgarian, which can be used to enrich the ICD-10 disease classification. It is worth mentioning that we partially handle this problem by automatically generating a gazetteer of diseases from the source documents.

We further found that we needed to remove a number of identifier references from the ICD-10 names. In particular, we filtered out any disease names that looked like codes and were in parentheses, e.g., (J99.8*), (L40.5ї), (M45-M46ї, M48.-ї, M53-M54ї), (E10-E14ї с общ четвърти знак .4) and the like. We further filtered out some abbreviations that did not refer to the target disease but to molecular markers or abnormal proteins and other participants that can cause the disease. Here are some examples:

```
G36.0 Оптиконевромиелит [болест на Devic]

G36.1 Остър и подостър хеморагичен левкоенцефалит [болест на Hurst]
```

In order to increase the number of actual disease names for which we could provide ICD-10 identifiers and to create additional synonyms for some of the diseases, we reordered and selectively extracted some names in parentheses from the existing disease names in ICD-10 (rather than the description) of the disease in such cases. For example, we rewrote the two examples above as follows:

```
G36.0 Оптиконевромиелит

G36.0 болест на Devic

G36.1 Остър и подостър хеморагичен левкоенцефалит

G36.1 болест на Hurst
```

In order to automatically prepare the corpus for manual annotation of disease mentions, we created a GATE processing tool that implements a number of string distance metrics based on *SimMetric*, an open source library of similarity and distance metrics, including Levenshtein, L2, Cosine, Jaccard, Jaro-Winkler, etc. *SimMetric* has a visual interface, which facilitates the selection of the most appropriate similarity measure for a particular task. After some preliminary experiments, we found Jaro-Winkler to be most fit for our data.

Based on the score of the distance match between a disease mention in the text and the names in the ICD-10 dictionary, we ordered and select the top-3 names from ICD-10. We then fed these top-3 candidates in a GATE user interface, specially created for the purpose, which is shown on Figure 2.

In Figure 2, the text and the diseases are shown in the upper part of the screen, while the disease names from ICD-10 with the top-3 Jaro-Winkler scores are shown in the bottom. A human annotator can delete the candidates that are incorrect in the given context with a single mouse click, e.g., "УЗД", which is a procedure/examination. In some rare cases, the annotator might also need to add new candidates, which can be done with a right click: see the case of "тиреоид на Хашимото", where the correct candidate is "E06.3 Автоимунен тиреоидит" instead of "E06.9 Тиреоидит, неуточнен" that is present in the top 3 candidates.

7 Conclusion and Future Work

In this work we focus on simple approaches to named entity recognition having limited or no prior example data. In this framework, we have demonstrated that a seemingly complicated named entity recognition task can be handled with satisfying quality in a fast and robust manner. We have achieved this by examining the structure and language expressions, as well as words and orthographic features found in clinical epicrises. We have further demonstrated that using general purpose frameworks such as GATE and an appropriate pragmatic methodology can significantly speed up the process of annotation.

Our disease mentions recognizer, annotation guidelines and annotated epicrises are potentially useful for applications such as document categorization and search. Moreover, extending the disease mention recognition to semantic annotations with identifiers from an ontology such as ICD-10 would enable a number of applications such as monitoring disease progression and manifestation over a period of time and linking epicrises to a web of linked data like *LinkedLifeData*. We believe these are promising research directions and we plan to pursue them in future work.

For the purpose of facilitating and speeding up manual semantic annotations, we have developed a new GATE-based processing tool that can calculate string similarity scores between disease mentions found in text and disease names listed in ICD-10. We have further coupled this with a GATE visual resource that allows a human annotator to delete wrong mentions at a given offset, thus only leaving the correct option(s), while also allowing the addition of more options. In future work, we plan to use this interface in a similar pragmatic approach to the task of normalizing disease names in context with respect to ICD-10.

Acknowledgements

The research presented in this paper is partially supported by grant DO 02-292/December 2008 funded by the Bulgarian National Science Fund.

We would like to thank the anonymous reviewers for their detailed and constructive comments, which have helped us improve the paper.

References

Hamish Cunningham. 2000. *Software Architecture for Language Engineering*. PhD thesis, University of Sheffield, Sheffield, UK.

Marjorie Freedman, Lance Ramshaw, Elizabeth Boschee, Ryan Gabbard, Gary Kratkiewicz, Nicolas Ward and Ralph Weischedel. 2011. Extreme Extraction — Machine Reading in a Week. *Proceedings of the Conference on Empirical Methods in Natural Language Processing (EMNLP'2011)*, pp. 1437–1446, Edinburgh, Scotland, UK.

Yoav Freund and Robert Schapire. 1999. Large margin classification using the perceptron algorithm. *Machine Learning*, 37(3):277-296.

Kuzman Ganchev, Fernando Pereira, Mark Mandel, Steven Carroll and Peter White. Semi-Automated Named Entity Annotation. 2007. In *Proceedings of the Linguistic Annotation Workshop (LAW'07)*. pp. 53-56, Prague, Czech Republic.

Georgi Georgiev, Preslav Nakov, Kuzman Ganchev, Petya Osenova, Kiril Simov. 2009. Feature-Rich Named Entity Recognition for Bulgarian Using Conditional Random Fields. In *Proceedings of the Conference on Recent Advances in Natural Language Processing (RANLP'2009)*. pp. 113-117, Borovets, Bulgaria.

Ryan McDonald and Fernando Pereira. 2005. Identifying gene and protein mentions in text using conditional random fields, *BMC Bioinformatics*, 6(Suppl 1):S6 doi:10.1186/1471-2105-6-S1-S6

Douglas W. Oard. 2003. The Surprise Language Exercises. *ACM Transactions on Asian Language Processing*, 2(2):79–84.

Burr Settles. 2011. Closing the Loop: Fast, Interactive Semi-Supervised Annotation With Queries on Features and Instances. *Proceedings of the 2011 Conference on Empirical Methods in Natural Language Processing (EMNLP'2011)*, pp. 27–31, Edinburgh, Scotland, UK.

Koby Crammer, Mark Dredze, Kuzman Ganchev, Pratim Partha Talukdar and Steven Carroll. 2007. Automatic Code Assignment to Medical Text. *In Workshop on biological, translational, and clinical language processing (BioNLP)*, pp. 129-136, Prague, Czech Republic.

Reducing Complexity in Parsing Scientific Medical Data, *a Diabetes Case Study*

Dimitrios Kokkinakis
Center for Language Technology and Språkbanken
Department of Swedish
University of Gothenburg, Sweden
dimitrios.kokkinakis@svenska.gu.se

Abstract

The aim of this study is to assemble and deploy various NLP components and resources in order to parse scientific medical text data and evaluate the degree in which these resources contribute to the overall parsing performance. With *parsing* we limit our efforts to the identification of unrestricted noun phrases with full phrase structure and investigate the effects of using layers of semantic annotations prior to parsing. Scientific medical texts exhibit complex linguistic structure but also regularities that can be captured by pre-processing the texts with specialized semantically-aware tools. Our results show evidence of improved performance while the complexity of parsing is reduced. Parsed scientific texts and inferred syntactic information can be leveraged to improve the accuracy of higher-level tasks such as information extraction and enhance the acquisition of semantic relations and events.

1 Introduction

Linguistic annotation of textual corpora in any field, and in specialized fields in particular, is a demanding and complex task, absolute necessary for data-driven language processing, human language technologies and knowledge mining. One such type of processing is at the syntactic level, i.e. *syntactic parsing*. The aim of this study is to develop and evaluate a method of identifying unrestricted noun phrases with full phrase structure from a scientific medical corpus. To ease the evaluation, in lack of an appropriate gold standard, we selected random sentences from the available corpus with mentions of the word *diabetes*. This subset then was automatically annotated and manually inspected and corrected. Furthermore, we tried to be minimalistic by assembling and deploying various *existing* NLP components and resources in order to evaluate the degree in which these resources contribute to the overall parser's performance. Analysis of scientific texts is a challenging task caused by deviant and idiosyncratic uses of vocabulary and syntax and complex linguistic structure. However, we believe that there are also regularities that can be captured by pre-processing the texts with specialized semantic tools. This way the complexity of parsing in scientific discourse, e.g., ambiguities, can be reduced, while the technical vocabulary increases the lexical coverage. For parsing we use finite-state cascades and sequential finite-state transducers. The focus of the current work is on the extraction of complete noun phrases, an important step that upon succession paves the way for the extraction of more complex structures and functional syntactic relations. Parsing is important for in-depth semantic interpretation; inferred syntactic and semantic information can be used to improve the accuracy of higher-level tasks such as information extraction and enhance the acquisition of relations and events.

2 Background

Parsing technology has seen a dramatic improvement over the last decade and a number of fairly robust parsers are available for a growing number of languages and application domains. This is a trend that has been accelerated by the appearance of wide coverage grammars and statistical parsing modules, both based on the availability of various treebanks such as the Penn or the GENIA treebanks (*cf.* Rimell & Clark, 2009). The commonest strategies to parsing are constituency/phrase structure or dependency parsing; for a review of parsing strategies, *cf.* Ljunglöf & Wirén (2010). In the first, words combine into phrases which repeatedly combine to form the sentence; while in the second syntactic analysis take the form of binary relations, that hold between words; Pyysalo (2008); Nivre (2005). Phrase structure grammars yield fast and reliable results without the need of large (annotated) corpora while dependency parsing is essentially

Proceedings of the Workshop on Biomedical Natural Language Processing, pages 35–42,
Hissar, Bulgaria, 15 September 2011.

very similar to the concept of *valency* extended to all word classes. In the medical field, there have been a number of approaches to syntactic parsing (Leroy *et al.*, 2003; Ohta *et al.*, 2005; Lease & Charniak, 2005). The goal for most of these approaches was with the extraction of various types of relations between phrases with named entities, e.g. proteins, since good precision and recall figures for extracting such relations requires a reliable syntactic analysis of the text. A large body of work to dependency parsing in the (bio)medical domain is based on the GENIA corpus Kim *et al.* (2003); see for instance Rinaldi *et al.* (2008) and Pyysalo (2008). Nevertheless, the issue of domain adaptation of *existing* grammars is still an open issue. It has been discussed that adaptation efforts should be on lower, local levels of representation (domain specific part-of-speech, dictionary collocations, named entities, terminology) not on full parse trees (Leash & Charniak, 2005; Huang *et al.*, 2005; Aubin *et al.*, 2005; Grover *et al.*, 2005; Hogan *et al.*, 2011). For example, Leash & Charniak (2005) showed clear improvements of the parsing accuracy considering a combination part-of-speech/named entities. Accuracy increased from 81.5% to 82.9% of a Penn Treebank-trained parser applied on biomedical literature. For a survey of comparing and combining six state-of-the-art *chunkers* for the biomedical domain see Kang *et al.*, (2010).

3 Materials and Method

The ever-increasing amount of biomedical (molecular biology, genetics, proteomics) and clinical data repositories increase in a dramatic manner. Such data appropriately annotated with event-level information are a valuable source of evidence-based research and text mining activities, such as information extraction, semantic search, question&answering and knowledge discovery. Syntactic parsing is considered an important ingredient for *event-based information extraction* from medical free text. Extracting pieces of information pertaining to specific events requires the extraction of argument mentions, often syntactic, that play a specific role within the event. In order to support the automated extraction of events, annotated corpora with event-level information is a necessary requirement; *cf.* Wattarujeekrit *et al.* (2004) and Thompson *et al.* (2009).

For our study we have selected a random sample of 120 sentences[1] with the mention of the word *diabetes* from a large corpus of scientific medical Swedish (Kokkinakis & Gerdin, 2010). The average length of a tokenized sentence in the sample is 23,8 tokens. Despite the small size of the sample, we can still find characteristics, typical of the medical scientific language such as terminology overload and coordinative constructions. For instance, it is rather common with coordinative phrases such as: *Man har visat att serumnivåerna av CRP, IL-6, fibrinogen, PAI-1, amyloid A och sialinsyra är förhöjda vid typ 2-diabetes* 'It has been shown that serum levels of CRP, IL-6, fibrinogen, PAI-1, amyloid A and sialic acid are elevated in type 2 diabetes'. These characteristics are expressed by a syntactic and vocabulary variability which compared to "ordinary" language requires adapted parsing strategies in order to be able to effectively capture the peculiarities of the genre. Terminology and named entity recognition can contribute to reliably resolve some of these problems.

3.1 Parsing Method

For the identification and labeling of noun phrases we apply an *easy-first parsing* deterministic approach using finite-state cascades. A finite-state cascade consists of a sequence of levels; phrases at one level are built on phrases at the previous level. Levels consist of rules, *alias* groups of patterns, ordered according to their internal complexity and length. A pattern consists of a category and a regular expression and parsing consists of a series of finite transductions. Spans of input elements are reduced to a single element in each transduction; i.e. regular expressions are translated into finite-state automata, the union of which yields a single, deterministic, finite-state, level recognizer; for further details of the approach *cf.* Abney (1997). We use an existing generic grammar for modern Swedish which has been evaluated for both basic noun phrases and functional labels in general language corpora, for the noun phrases the precision reported was 97.82% and recall 94.5% (but *without* resolved attachments); details are reported in Kokkinakis & Johansson Kokkinakis (1999). The workflow of all the various steps are shown in Figure 1; here "attachment" refers to nouns and adjectives while the "term & entity aware rules" are integrated in the parser.

[1]The sample and some of the resources, e.g. multiwords, can be found at: <http://demo.spraakdata.gu.se/svedk/parse/>.

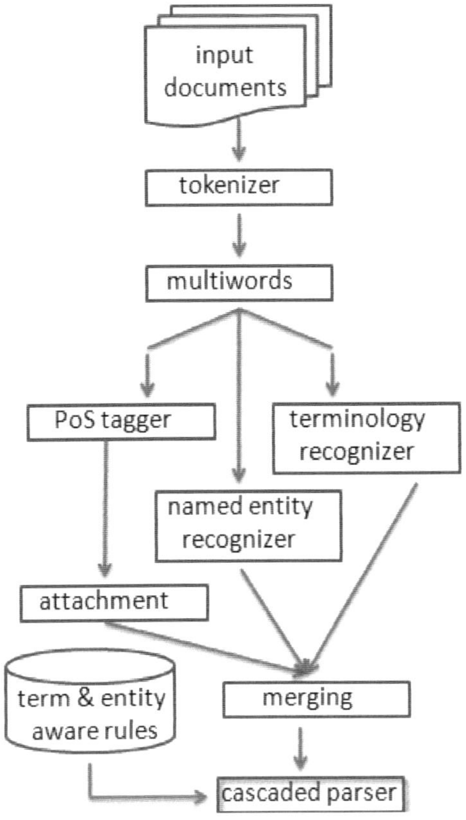

Figure 1. The workflow of the parsing process.

3.2 Parsing Adaptation, *Preliminaries*

Manual annotation of large amount of data with complex linguistic information, such as syntactic trees, is a costly enterprise, and means to remedy for this should be exploited. Moreover, since many parsers rely on several layers of representation there are various possible ways to enhance their performance even in different domains than the one they have been designed for. Our methodology is motivated by the fact that parsing performance can be gained by applying and improving on a number of pre-parsing stages. The idea is that various morphosyntactic and semantic representation layers can pave the way of substantial text complexity reduction as long as the parser can be made *aware* of these layers. Therefore, by putting effort on various levels of representation (pre-processing) we can hypothesize, and actually show, that performance can be improved. Thus, we help the parser in such a way that it can avoid some hard decisions, e.g. bracketing and structural ambiguities. We follow, and to a certain degree, extend the idea of Lease & Charniak (2005) discussed earlier. Our strategy is primarily based on four (domain) adaptations:

- *recognition of multiword expressions*
- *recognition of medical terminology*
- *recognition of named entities*
- *attachment for nouns & adjectives*.

In an indirect way, the recognition of terminology and named entities implies that multiword expressions are also recognized and the number of unknown words is reduced, while the lexical coverage increases. Consider for instance the examples: *Drottning Silvias Barn- och Ungdomssjukhus* 'Queen Silvia Children's Hospital' in which 5 tokens constitute a *coherent* entity and the more complex: *DIGAMI 1-studien (Diabetes mellitus insulin glucose infusion in acute myocardial infarction 1) [3] visade att [...].* 'The DIGAMI-1 study [...] showed that [...]' in which 9 tokens constitute a single entity. Apart from the terminology recognition (and the manual addition of domain vocabulary at the lexical resources we use) the rest are domain-independent adaptations.

3.3 Parsing Adaptation, *Steps*

Adaptations deal with the resolvement of at least *some* of the possible types of problems that can arise during parsing. For instance, we manage to dramatically increase the lexical coverage by efficiently dealing with *unknown words*, e.g., genre specific vocabulary. The majority of unknown words can be captured by the use of domain terminologies. A number of individual terms from such terminologies have been incorporated into the part-of-speech tagger's lexicon, for that purpose we use the TnT tagger (Brants, 2000). Similarly, for various types of multiword tokens, we have manually added a large number of common multiword function words (e.g., adverbs, preposition, determiners) in the part-of-speech tagger's lexicon[2]. While for the majority of other types of multiword expressions (i.e., terms and named entities) which are identified during terminology and named entity recognition, possibly erroneous part-of-speech annotation does not have impact during parsing. For instance the part-of-speech annotation of the segment: *en latent diabetes mellitus* 'a latent diabetes mellitus' becomes: *en/DI@US@S latent/AQPUSNIS diabetes/NCUSN@IS mellitus/XF* (XF stands here as a tag for foreign

[2] Nivre & Nilsson (2004) have showed that significant improvement in parsing accuracy for Swedish could be achieved if multiword function words are taken under consideration.

words) while the parsing of this segment (in a simplified form) becomes ***np:*** *<en latent diabetes>* ***np:*** *<mellitus>*, that is two separate noun phrases. However, if we apply terminology recognition and then combine (in some suitable way) that information with the part-of-speech (e.g., by adding a feature to the part-of-speech) then we end with the following annotation: *en/DI@US@S latent/AQPUSNIS diabetes/NCUSN@IS-**TRM-B** mellitus/XF-**TRM-I***. The parser, being aware of these new features, will treat *diabetes* and *mellitus* as a unit (e.g., with a rule such as ART? ADJ* TERM+) and *favorize* the term annotation, since term and entity label features are given higher precedence compared to part-of-speech tags by the parser. In this case, the parser produces a correct phrasal constituent, one noun phrase, namely ***np-mdcn:*** *<en latent diabetes mellitus>*.

The following example will be used to illustrate some of these steps in sections 3.3.1-3.3.3:

Malmö 22 januari 2008 - Nya data som publiceras idag styrker effektiviteten hos basinsulinet Levemir® (insulin detemir) som behandling en gång om dagen för personer med typ 2-diabetes.

Lit: "Malmö 22 January 2008 - New data published today, confirm the effectiveness of basal insulin Levemir® (insulin detemir) as a treatment once a day for people with type 2 diabetes.".

3.3.1 Medical Terminology Recognition

We use the Swedish **S**ystematized **No**menclature of **Med**icine, **C**linical **T**erms (SNOMED CT) for terminology recognition. Terminology is actually used for two reasons: (i) to improve the performance of the generic part-of-speech tagger and (ii) to actually aid the recognition of the terminology and consequently also the annotation of terms in text in which the parser has been modified to be aware of. First, we extracted one-word terms (ca 30k) and semi-automatically added those with their full morphosyntactic description, to the part-of-speech tagger's *backup* lexicon. Using regular expressions over the suffixes of the terms we automatically added appropriate morphosyntactic descriptions and manually reviewed a number of unmatched cases, usually Latin terms, which we added with the label for *foreign words*. Secondly, we performed terminology recognition and then merged the output to the representation format required by the parser, thus the parser becomes aware of the terminology in a single, simple step. The previously mentioned example follows below after terminology annotation (annotations are given between the XML tag *snomed* with attributes *c* concept, *h* id-number and *o* original form). Note that for simplicity reasons *qualifier values* have been filtered away:

Malmö 22 januari 2008 - Nya data som publiceras idag styrker effektiviteten hos **<snomed c="substance" h="25305005" o="långtidsverkande insulin">***basinsulinet***</snomed>** *Levemir® (***<snomed c= "substance" h="414515005" o="detemir insulin">***insulin detemir***</snomed>***) som behandling en gång om dagen för personer med* **<snomed c="disorder" h="44054006" o="diabetes mellitus typ 2">***typ 2-diabetes***</snomed>.**

3.3.2 Named Entity Recognition

In exactly the same manner, as previously, we apply the generic named entity recognizer which also serves two important purposes. Firstly, to aid the recognition and annotation of single and multiword named entities and secondly, in an indirect way, to aid the appropriate recognition of (unknown) multiword expressions/tokens. The annotation of the previous example shown this time below illustrates how this type of annotation looks like. After named entity recognition the example sentence takes the following form:

<ENAMEX TYPE="LOC" SBT="PPL">*Malmö* **</ENAMEX> <TIMEX TYPE="TME" SBT="DAT">** *22 januari 2008***</TIMEX>** *- Nya data som publiceras* **<TIMEX TYPE="TME" SBT="DAT">** *idag***</TIMEX>** *styrker effektiviteten hos basinsulinet* **<ENAMEX TYPE= "OBJ" SBT= "MDC">***Levemir®***</ENAMEX>** *(insulin detemir) som behandling* **<NUMEX TYPE="MSR" SBT="FRQ">***en gång om dagen***</NUMEX>** *för* **<ENAMEX TYPE="PRS" SBT="CLC">***personer* **</ENAMEX>** *med typ 2-diabetes.*

In the above annotations, *ENAMEX* stands for a named entity, *TIMEX* for a time entity and *NUMEX* for a measure entity. All annotations produce also two attributes (not used in the current study) namely main *TYPE* and *SuBType*; details are provided in Kokkinakis (2004).

3.3.3 Structural Ambiguity / Attachment

For the structural ambiguity/preposition attachment disambiguation we use a generic Swedish valency/subcategorization lexicon which has been manually enhanced for genre-specific nouns (such as *ulceration*), which all take a contextually optional prepositional phrase as complement; e.g., *ulceration* ***av*** *tumören* 'ulceration of the tumor') and adjectives (such as *resistent*) which also take a contextually optional prepositional phrase as complement, e.g. *resistent* ***mot*** *autokrint insulin* 'resistant to autocrine insulin'. There is also a small number of nouns that show a semantic preference for two arguments, such as *övergång* 'transition, as in *övergång* ***från*** *blodglukos* ***till*** *plasma-glukos* 'transition from blood-glucose to plasma glucose'. This type of lexical information is applied after part-of-speech tagging using a contextually-driven filter that determines whether a suitable feature can be added to nouns' or adjective's part-of-speech annotation. This is a naïve but reliable way to capture lexical semantic preferences without a lot of effort. Thus, in the example from section 3.3 there are two such tokens identified and annotated with the feature *-VAL*, for *valency* (attached to the appropriate nominal or adjectival heads); the morphosyntactic descriptions *NCUSN@DS*, *NCUSN@IS* stand for common nouns and *SPS* stands for a preposition, the tags for the rest of the words have been omitted for simplicity. The tagset we use is an extended version of the Swedish MULTEXT tagset[3].

*Malmö 22 januari 2008 - Nya data som publiceras idag styrker effektiviteten/NCUSN@DS-**VAL** **hos/SPS** basinsulinet Levemir® (insulin detemir) som behandling/NCUSN@IS-**VAL** en gång om dagen* ***för/SPS*** *personer med typ 2-diabetes.*

3.3.2 Merging and Parsing Awareness

All results are merged into a uniform representation. In order to make the parser aware of all the annotations we have added two new *levels* of manually written rules into the parser's original sequence of levels. Recall that each level contains a handful of rules, and phrases recognized by the rules of one level are built on phrases at the previous level. The two new levels, at the very beginning of the parser's level set, are used to *only* recognize and process the sequences of terminology and named entity annotations, labeling them as either noun phrases, e.g., *np-location, np-disorder*, or adverbial phrases, as in the case of time expressions. Examples of rules for these new levels and an example of parsing output are given in Appendix A&B. The parser provides several possible ways to produce output results. For instance, all features can be explicitly generated and in the example we can see *lemma*, the base form of each token and *sem* the semantics (term or entity labels) or *N/A* (non-applicable) otherwise.

Note that each level is proceeded by abbreviated bundles of enhanced part-of-speech tags, e.g., one mnemonic name for all adjectives (*ADJ*) or one for all possible location annotations (*LOC-B* and *LOC-I*). By this technique the actual rules become simpler, flexible and human readable; examples are also given in Appendix B. This convention does not exclude the possibility to actually use a particular part-of-speech tag or even word in a rule, which implies that rules can be *lexicalized* i.e. use words in the production rules.

4 Results and Evaluation

We performed a four-stage evaluation in order to measure the contribution of the adaptations to the overall performance of the results.

Model	**Pr**	**R**	**F-m**	**F-impr.**
baseline *comp. to the gold*	39.53%	47.22%	43.04%	
+backup+mwe	56.76%	58.33%	57.53%	**14,4%**
+backup+mwe+val	53.66%	61.11%	57.14%	**14.1%**
+backup+mwe+val+NER	64.10%	69.44%	66.67%	**23,6%**
+backup+mwe+val+NER+term	89.19%	91.67%	90.41%	**47,3%**

Table 1: Evaluation results

We manually corrected the output of the automatic parsing with *all* adaptations and used it as a gold standard for the evaluation, using the 2008-version of the *evalb* software by Sekine & Collins with default values. The only change we made was to convert the parsing output in order to run *evalb*. Brackets, that the parser returns as delimiters, were converted to parentheses, that *evalb* requires. The results in table 1 stand for: *Pr* bracketing precision, *R* bracketing recall, *F* bracketing f-measure and *F-impr.* for f-measure improvement.

We also measured the number of unknown words in the sample before and after the vocabulary enhancement of the part-of-speech tagger. The results from the part-of-speech tagging for the sentences examined showed only a small improvement, mainly because the majority of unknown words were actually annotated correctly by the tagger; 388 tokens (13,6%) were unknown (5,4% of those were wrongly annotated), reduced to 215 unknown tokens (7,5%) with the use of the enhanced vocabulary. The number of multiword function words was 27, e.g., *t ex* 'for example'.

[3] <http://spraakbanken.gu.se/parole/tags.phtml> (last visited in August 2011).

5 Conclusions

To our knowledge there hasn't been any other report to use of parsing Swedish medical corpus of any type. Therefore any direct comparisons are difficult to make. With simple modifications, as shown in Table 1, the parser becomes aware of the different (shallow) semantic annotations produced during pre-processing. We believe of course that any type of parsing strategy can benefit from the integration of this type of annotations. The results show a substantial improvement of accuracy which can be attributed to a number of factors such as structural ambiguity reduction, increasing lexical coverage, enhanced processing of coordinative structures. For the future we intend to extend the subcategorizations to verbs, particularly since relevant Swedish lexical resources are available for that purpose. We also plan to adapt the rest of the original, generic parser in order to be able to support event-based information extraction from Swedish medical corpora.

Acknowledgments

This work is partially supported by the Centre for Language Technology (CLT) <http://clt.gu.se/>.

References

Abney S. 1997. Part-of-Speech Tagging and Partial Parsing. In *Corpus-Based Methods in Language and Speech Processing*. Young & Bloothooft (eds). 4:118-136. Kluwer.

Aubin S., Nazarenko A. and Nédellec C. 2005. Adapting a general parser to a sublanguage. Recent Advances in Natural Language Processing (RANLP) Pp. 89-93. Bulgaria.

Brants T. 2000. TnT: a statistical part-of-speech tagger. Sixth Conference on Applied NLP. Seattle, USA.

Grover C., Lapata M. and Lascarides A. 2005.A Comparison of Parsing Technologies for the Biomedical Domain. *Natural Language Engineering* 11(1), 27-65, CUP.

Hogan D., Foster J. and van Genabith J. 2011. Decreasing Lexical Data Sparsity in Statistical Syntactic Parsing - Experiments with Named Entities. Proceedings of the ACL Workshop: Multiword Expressions: from Parsing and Generation to the Real World (MWE). Portland, USA

Huang Y., Lowe HJ., Klein D. and Cucina RJ. 2005. Improved Identification of Noun Phrases in Clinical Radiology Reports.. *J Am Med Inform Assoc.* 12(3): 275–285.

Kang N., van Mulligen EM. and Kors JA. 2010. Comparing and combining chunkers of biomedical text. *J Biomed Inform.* 44(2):354-60. Epub Nov 4.

Kim J-D., Ohta T. Tateisi Y. and Tsujii, J. 2003. GENIA corpus - a semantically annotated corpus for bio-textmining. *Bioinformatics*. 19 Suppl 1:i180-2.

Kokkinakis D. and Johansson Kokkinakis S. 1999. A Cascaded Finite-State Parser for Syntactic Analysis of Swedish. 9th European Chapter of the Association of Computational Linguistics (EACL). 245-248. Norway.

Kokkinakis D. 2004. *Reducing the Effect of Name Explosion.* Workshop: Beyond Named Entity Recognition, Semantic Labelling for NLP tasks. In conjunction with the 4th Language Resources and Evaluation Conference (LREC). Lisbon, Portugal.

Kokkinakis D. and Gerdin U. 2010. A Swedish Scientific Medical Corpus for Terminology Management and Linguistic Exploration. Seventh Language Resources and Evaluation Conference (LREC). Malta.

Lease M. and Charniak E. 2005. Parsing Biomedical Literature. Second International Joint Conference on Natural Language Processing. 58-69. Korea.

Leroy G., Chen H. and Martinez J.D. 2003. A shallow parser based on closed-class words to capture relations in biomedical text. *J Biomed Inform.* 36(3):145-58.

Ljunglöf P. and Wirén M. 2010. Syntactic Parsing. In *Handbook of Natural Language Processing*, 2nd edition. Indurkhya N. and Damerau FJ. (eds). Pp. 59-91. CRC Press.

Nivre J. 2005. Dependency Grammar and Dependency Parsing. Technical Report. Växjö University.

Nivre J. and Nilsson, J. 2004, Multiword Units in Syntactic Parsing. MEMURA 2004 workshop. pp. 39-46. Lisbon

Ohta T., Tateisi Y. and Tsujii J. 2005. Syntax Annotation for the GENIA corpus. IJCNLP, pp. 222-227. Korea.

Pyysalo S. 2008. *A Dependency Parsing Approach to Biomedical Text Mining.* Turku Centre for Computer Science. TUCS Dissertations: 105, U of Turku, Finland.

Rimell L. and Clark S. 2009. Porting a lexicalized-grammar parser to the biomedical domain. *J of Biomed Inf.* 42:5.

Rinaldi F., Schneider G., Kaljurand K. and Hess, M. 2008. Dependency-Based Relation Mining for Biomedical Literature. Sixth Language Resources and Evaluation Conference (LREC). Morocco.

Sekine S. and Collins MJ. 2008. EVALB. <http://nlp.cs.nyu.edu/evalb/EVALB20080701.tgz>

Thompson P., Iqbal SA. McNaught J. and Ananiadou S. 2009. Construction of an annotated corpus to support biomedical information extraction. *BMC Bioinformatics*, 10:349.

Wattarujeekrit T, Shah PK. and Collier N. 2004. PASBio: predicate-argument structures for event extraction in molecular biology. *BMC Bioinformatics.* 19;5:155.

APPENDIX A

```
<s id="955">
  [np-location
    [NPOON@OS-LOC-B <t id="955_1"/> Malmö lem=malmö]]
  [rp-time
    [MCOONOS-TME-B <t id="955_2"/> 22 lem=22]
    [NCUSN@IS-TME-I <t id="955_3"/> januari lem=januari]
    [MCOONOS-TME-I <t id="955_4"/> 2008 lem=2008]]
  [FI <t id="955_5"/> - lem=-]
  [np
    [AQPOPNOS <t id="955_6"/> Nya lem=ny]
    [NCNPN@IS <t id="955_7"/> data lem=data]]
  [np-som
    [PH@OOO@S <t id="955_8"/> som lem=som]]
  [vg_p_f
    [V@IPSS <t id="955_9"/> publiceras lem=publicera]]
  [rp-time
    [RGOS-TME-B <t id="955_10"/> idag lem=idag]]
  [vg_a_f
    [V@IPAS <t id="955_11"/> styrker lem=styrka]]
  [np_attach_pp
    [np-val
      [NCUSN@DS-VAL <t id="955_12"/> effektiviteten lem=effektivitet]]
    [pp
      [SPS <t id="955_13"/> hos lem=hos]
      [np-medical
        [NCNSN@DS-MDC-B <t id="955_14"/> basinsulinet lem=basinsulin]
        [NPOON@OS-MDC-B <t id="955_15"/> Levemir® lem=levemir®]
        [FP <t id="955_16"/> ( lem=(]
        [NCNSN@IS-MDC-B <t id="955_17"/> insulin lem=insulin]
        [XF-MDC-I <t id="955_18"/> detemir lem=detemir]
        [FP <t id="955_19"/> ) lem=)]]]]
  [CCS <t id="955_20"/> som lem=som]
  [np_attach_pp
    [np-val
      [NCUSN@IS-VAL <t id="955_21"/> behandling lem=behandling]
       [np-mesr-frq
         [DI@US@S-MSR-B <t id="955_22"/> en lem=en]
         [NCUSN@IS-MSR-I <t id="955_23"/> gång lem=gång]
         [SPS-MSR-I <t id="955_24"/> om lem=om]
         [NCUSN@DS-MSR-I <t id="955_25"/> dagen lem=dag]]]
    [pp
      [SPS <t id="955_26"/> för lem=för]
      [np-person
        [NCUPN@IS-PRS-B-VAL <t id="955_27"/> personer lem=person]]]
    [pp
      [SPS <t id="955_28"/> med lem=med]
      [np-medical
        [NCUSN@IS-MDC-B <t id="955_29"/> typ lem=typ]
        [NCUSN@IS-MDC-I <t id="955_30"/> 2-diabetes lem=2-diabetes]]]]
  [FE <t id="955_31"/> . lem=.]
</s>
```

Example parsing with basic constituent annotations including the part-of-speech and the feature *lem*[ma].

APPENDIX B

```
#Level 'medical entities'
:mdcn-entity
 article = DO@OP@S|DO@US@S|DO@OP@S|DO@NS@S|DF@OP@S|DF@OS@S|DF@NS@S|DF@US@S|...;
 adjective = AFOOPGOS|AFOOPNOS|AFOOSNDS|AFOMSNDS|AFONSNIS|AFOUSNIS|APOOONOS|...;
 adverb = RGOA|RGOC|RGOS|RGCS|RGPS|RGSS|RHOS;
 numerical = MCOOOOC|MCOOGOS|MCOONOS;
 MDCN-B = AFOOOOOA-MDC-B|APOOONOS-MDC-B|AFOOPGOS-MDC-B|AFOOPNOS-MDC-B|...;
 MDCN-I = AFOOOOOA-MDC-I|APOOONOS-MDC-I|AFOOPGOS-MDC-I|AFOOPNOS-MDC-I|...;

 np-medical ->  article? (numerical|adjective)* MDCN-B MDCN-I*
              # Rules for Special Cases:
              #      e.g.  'vita "leverfläckar" i.e., 'while "liver spots"'
              | article  (numerical|adjective|adverb)* FP MDCN-B MDCN-I* FP
              | article? FP (numerical|adjective|adverb)+ FP MDCN-B MDCN-I*
              | ...
;
#Level 'location entities'
:loc-entity
 article = DO@OP@S|DO@US@S|DO@OP@S|DO@NS@S|DF@OP@S|DF@OS@S|DF@NS@S|DF@US@S|...;
 adjective = AFOOPGOS|AFOOPNOS|AFOOSNDS|AFOMSNDS|AFONSNIS|AFOUSNIS|APOOONOS|...;
 adverb = RGOA|RGOC|RGOS|RGCS|RGPS|RGSS|RHOS;
 numerical = MCOOOOC|MCOOGOS|MCOONOS;
 LOC-B = AFOOOOOA-LOC-B|APOOONOS-LOC-B|AFOOPGOS-LOC-B|AFOOPNOS-LOC-B|...;
 LOC-I = AFOOOOOA-LOC-I|APOOONOS-LOC-I|AFOOPGOS-LOC-I|AFOOPNOS-LOC-I|...;

 np-location -> article? (numerical|adjective)* LOC-B LOC-I*
              # Rules for Special Cases, e.g.  'USA, Canada och England'
              | LOC-B (FI LOC-B)+ CCS LOC-B
              | article  (numerical|adjective|adverb)* FP LOC-B LOC-I* FP
              | article? FP (numerical|adjective|adverb)+ FP LOC-B LOC-I*
              | ...
;
#Rest of the grammar ...
...
#Level 'attachment'
:attachment
np = np-medical|np-location|...;
np-val = NCUSN@DS-VAL|NCUSN@IS-VAL|NCNPN@IS-VAL|...;

# Attach a PP using the 'VAL' feature that is added to certain tokens after
#   part-of-speech tagging
 np_attach_pp ->
 # Here, 'SPS' is the part-of-speech for any preposition
       np-val [pp = SPS np]
;
```

Part of the grammar that shows several available levels. The first two (and new to the existing generic grammar) deal with medical and location entities followed by other entity specific and then by general rules (not shown here). At the end of the figure there is an example of an attachment rule, applied after the recognition of basic phrase constituents, e.g. noun phrases and verbal groups (i.e., an obligatory lexical head plus optional auxilaries or even adverbs if those intervene between an auxiliary and a head verb).

Architecture and Systems for Monitoring Hospital Acquired Infections inside a Hospital Information Workflow

Denys Proux[1], Caroline Hagège[1], Quentin Gicquel[2], Suzanne Pereira[3], Stefan Darmoni[4], Frédérique Segond[1], Marie-Hélène Metzger[2]

(1) XRCE, 6 Chemin de Maupertuis, 38240 Meylan, France
(2) UCBL-CNRS, UMR 5558 Lyon, France
(3) CISMEF, Rouen, France
(4) VIDAL, Issy les Moulineaux, France
Denys.Proux@xrce.xerox.com, Caroline.Hagege@xrce.xerox.com, Quentin.Gicquel@chu-lyon.fr, Suzanne.Pereira@vidal.fr, Stefan.Darmoni@cismef.fr, Frederique.Segond@xrce.xerox.com, Marie-Helene.Metzger@chu-lyon.fr

Abstract

This paper describes the latest developments in the design of a tool to monitor Patient Discharge Summaries to detected pieces of evidences related to Hospital Acquired Infections. Anonymization, Named Entity detection, Temporal Expressions analysis and Causality detection methods have been developed and evaluated. They are embedded in a tool designed to work in a Hospital Information Workflow.

1 Information and Communication Technologies to Improve Patient Safety

Managing information related to Patient Records (PR) is something complex. Treating a patient is not like fixing a tire; it is a long process that involves many medical disciplines. For each analysis, each treatment, each diagnosis, fragmented information is produced by different people, different medical units. Information and Communication Technologies (ICT) appears to be a good opportunity to make use of this information to offer monitoring and alert services which contribute in the end to patient safety. Among these opportunities Hospital Acquired Infections is a domain where ICT can bring a lot to help experts.

1.1 The problem

Hospital Acquired Infections (HAI) can be defined as: *An infection occurring in a patient in a hospital or other health care facility in whom the infection was not present or incubating at the time of admission. This includes infections acquired in the hospital but appearing after discharge, and also occupational infections among staff of the facility.* (Garner 1988). This problem occurs because hospitals are special places concentrating both weak people and various types of diseases and infections. Not all HAI have the same effect but they all jeopardize patient safety and increase the time spent in hospitals.

1.2 Attempt to address the problem

In order to address this issue several efforts have been made mainly through the creation of a strict health protocol for the medical staff and appropriate training provided to this staff. Furthermore experts have been appointed to monitor these risks and a strict reporting process has been setup. However these efforts are not completely successful or at least could be improved. This is mainly due to both the way information is managed inside hospitals, and the inherent complexity of HAI.

To perform their analysis experts need to have access to many data / reports to make a decision about the causality of an infection. This is difficult to obtain for experts not only because information is spread out into various databases but also because, as they are dealing with constantly evolving information. Therefore they need a tool to automate report monitoring. Such tool does not exist as one of the major issues is

Proceedings of the Workshop on Biomedical Natural Language Processing, pages 43–48,
Hissar, Bulgaria, 15 September 2011.

that information, most of the time, remains in an unstructured free text format.

This conclusion has pushed for the development of a research project bringing together HAI experts, medical terminology experts and Natural Language Processing experts to design a monitoring tool for patient records. The ultimate goal is to automate detection of HAI events from patient records.

2 A Natural Language Processing Approach to Monitor Patient Records

This 3 year project, started in January 2009, is conducted in collaboration with 3 University hospitals providing 1500 patient discharge summaries (half of them dealing with HAI). These reports have been manually analyzed 3 times by 3 different HAI experts to identify and annotate all pertinent elements. These elements have been indexed using a custom made tool connected with the FMTI multi-terminology server provided by project partners. These annotated documents are spitted into 3 sets used for designing the HAI detection rules, and to serve as a gold standard for 2 rounds of evaluation

2.1 Terminologies

Based on discussions between HAI experts and Terminology experts it has been decided (Metzger 2009) that only the following terminologies are necessary to reach our objective which is the detection of HAI events from reports:

- symptoms/diagnosis: CIM10, SNOMED3.5, MeSH
- bacteriological exams : SNOMED3.5, MeSH
- type of microorganisms : SNOMED3.5
- biological exams : SNOMED3.5, MeSH
- radiological exams : SNOMED 3.5, MeSH, CCAM
- Antibiotics : ATC, MeSH
- Type of surgical intervention: CCAM, MeSH

These terminologies are used to index pertinent named entities inside input reports. However after a first set of experiments we came to the conclusion that using as it is all the vocabulary contained in these terminologies generate noise and ambiguities for the detection of very specific information. Indeed, indexing is designed to maximize recall, but in our case we need to maximize precision. Therefore, based on the result of the annotation step we have decided to build our own HAI terminology which is a sub-part of these terminologies.

2.2 Event detection

The heart of this project is an Incremental Parser (XIP), which performs text mining. This parser is robust that is to say it has already been used in various projects to process large collections of unrestricted documents (web pages, news, encyclopedias, etc.) It has been designed to follow strict incremental strategies when applying parsing rules. The system never backtracks on rules to avoid falling into combinational explosion traps which makes it very appropriate to parse real long sentences from scientific texts for example (Aït-Mokhtar 1997). The analysis is relying on three processing layers which are: Part of Speech Disambiguation, Dependency Extractions between words on the basis of sub-tree patterns over chunk sequences, and a combination of those dependencies with Boolean operators to generate new dependencies or to modify or delete existing dependencies.

Named Entity detection and Event detection is performed using a standard French grammar that have been customized for medical language. As introduced in the previous section the issue with the scope of all selected terminologies forced us to develop and new terminology dedicated to HAI. To be more specific the system focuses on some very specific Named Entities and Events to perform the analysis. These elements are the following:

- Infectious germ: Bacteria, Virus, Yeast,
- Antiseptic : products used to clean or kill infectious germs
- Temperature : elements that indicate a fever or an abnormal change in body temperature
- Invasive devices: to perform measure or to cure. These devices can be an open gate for infectious germs.
- Exams: Such as bacteriologic, radiologic that could be the indication of a problem
- Treatments: all possible treatment that can be related or lead to an HAI (e.g. surgery)
- Diagnostic can also be used to take a decision (e.g. if it is explicitly said that an HAI occurred).

Furthermore, negation is also something to be taken into account for appropriate decision making. For example "no evolution on patient temperature" is a completely different statement compared to "the patient gets fever". Therefore appropriate negation management rules related to pertinent medical terms have been added to enrich the level of information extracted by the parser.

2.3 Temporality

HAI detection in Patient Discharge Summary (PDS) requires also an additional level of information to allow an accurate decision making process: temporality. HAI is not just about detecting isolated elements inside a report, it is also about matching the occurrence of these events with respect to a scenario. Detecting the right time stamp for these events is crucial as according to the time lap between two events (e.g. a knee surgery and an unexpected fever) is crucial to valid a possible HAI hypothesis.

Time detection rules have been designed and added to the parser. Time stamps are computed according to a reference date T0. The algorithm used for temporal indexing is detailed in (Hagège et. al., 2010).

2.4 Causality and decision heuristics

Among all challenges to detect an HAI event inside patient discharge summaries one of the most prominent is the fact that, most of the time, it does not appear explicitly inside texts. The only clue is a sequence of events occurring in a given time frame. This can be compared to a criminal investigation collecting pieces of evidence, searching for specific links between events, evaluating alibi, etc. Furthermore generally only few elements, separated in the text are present in the patient discharge summaries (Horan et al. 2008).

Therefore several discussions have brought together HAI experts and linguists to define which elements are necessary and what kind of relations are mandatory to come to a decision. There is a subtle difference between the official HAI definition and the type of information appearing in a patient record (PR).

A 1st set of heuristics have been designed to evaluate the ability of our system to detect HAI events inside patient records. These heuristics have been designed to maximize the recall. This means that not all the official rules have been encoded. These heuristics have therefore been created based on both formal definition and an empirical approach based on annotated samples.

To summarize this work, what is considered as a "smoking gun" in a patient records is:

- For Intensive Care Unit (ICU) at least one of the following criteria should be valid:
  - o If there is an explicit sentence speaking of a HAI
  - o If in close sentences we have at least 1 occurrence of both an Infection (e.g. germ) and an Antibiotic drug with time stamps at least equal to 2 days after T0 (T0+2), and no Infection event is described before T0 and the patient is alive.
  - o If the patient is already infected at T0 or if he has died during his stay and if at least 2 occurrences of either Infection, Antibiotic drug, Temperature, or an Invasive Device can be found with time stamps superior or equal to T0+2
- For a stay in an Surgery Unit at least one of the following criteria should be valid:
  - o If there is an explicit sentence speaking of a Surgery Site Infection
  - o If 1 of the following event can be detected with a time stamp superior to T0 : Infection, Antibiotic, Antiseptic, Germ, Bacteriological Exam.

These heuristics have been evaluated to estimate the level of improvement necessary to reach performance objectives expected by medical experts.

3 First Results

A preliminary experiment has been performed by our medical experts on 205 patient records. Results are presented in (Berrouane et al., 2011). The goal of this 1st experiment was to evaluate the efficiency and more specifically the recall of our heuristics to separate patient discharge summaries that deal with HAI and those that don't. On the evaluation corpus 128 patient records over 205 was dealing about HAI. The following table shows a brief overview of the results (details about the protocol are presented in (Hagege et al, 2011)). Here the recall is computed as True positive / (True Positive + False Negative), and Specificity is computed as True negative / (True Negative + False Positive).

For this experiment we compute Specificity instead of Precision (True Positive/(True Positive + False Positive)) as the distribution of the available corpus do not reflect the reality. In our corpus the number of positive and negative document are equal and the number of document per medical unit (Intensive Care Stomach Unit, Surgery, Orthopedic Surgery, Neuro-surgery) also do not reflect the same exact distribution as in a hospital.

	Patient Discharge Summaries	Recall	Specificity
All	205	87.6 %	97.4 %
ICU	29	62,5%	92.3%
Stomach Surgery	67	89,7%	100%
Orthopedic Surgery	21	87,5%	80%
Neuro-Surgery	88	93,1%	100%

Table 1: 1st results for automatic HAI detection

These results give only a flavor of the potential efficiency of the system. However this gives us good hope for the overall efficiency of the system as the global recall on our evaluation set reach 87.6 % with a Specificity of 97.4% before any improvement.

After some improvement a new experiment will take place at the end of 2011 on a final set of 800 Patient Discharge Summaries. But the success of this first evaluation campaign as pushed us to start developing an evolution of the prototype to plug it directly in a hospital information workflow for live evaluation.

4 Architecture for a Deployment in a real Hospital Information Workflow

The result of this 1st evaluation has demonstrated the potential of the detection system, however several assumptions have been made in the context of the research project and the evaluation is done on a set of ad-hoc documents prepared by medical experts participating to the project.

Therefore medical experts have asked for a special version of the system that could be directly plugged inside the hospital workflow to evaluate its performance in real life. Discussions have taken place to define the specifications and to prepare the delivery of such tool.

4.1 The patient record

After discussions it appears that the way information is managed inside an hospital information workflow is much more complex than simple collections of coherent patient discharge summaries. In fact in our case, each medical unit inside the hospital generates its own set of data when a treatment/analysis is performed. This information is both structured (for parts that can be structured) and unstructured (for free text comments, diagnosis, or summaries). However, even in a free text format, this information is always stored in text fields inside a database.

Furthermore patient information is very fragmented inside the database. Indeed, a patient can enter and leave the hospital several times in a given time frame, for different pathologies, and can travel across different medical units. This means that the global patient record evolve in time. So several questions have to be addressed:

- When the HAI detection system should be applied ?
- How to regroup coherent information related to a given patient (e.g. a left knee surgery, then 1 month later a right knee surgery, then after a new right knee surgery, etc)
- When can we decide that it is no more necessary to process new information?

In order to solve these questions we have defined with people managing the information system inside the hospital a specific architecture and HAI monitoring process for our system.

4.2 Architecture

It has been decided not to plug the HAI monitoring system directly inside the hospital information system (HIS) but rather to set it aside and to develop an ad-hoc standardization interface to allow further compatibility with potentially different types of hospital information systems. The process that is developed consists in:

- Each time new data is recorded inside the HIS for a given patient then a specific module is activated to gather all previous data recorded over a given time frame (currently over 1 year).
- A Custom Patient Record (CPR) is generated by the Data Gathering Module (DGM). This custom patient record is an XML document. Its structure is detailed in the next section. This document is pushed into a predefined temporary input repository.
- The HAI monitoring system browses the input repository and parses the content of

all custom patient records that are dropped in.
- If a HAI event is detected then related information is recorded in a specific database dedicated to HAI. This database allows expert to go back to the patient and to all documents they need to analyze the problem.

This architecture is designed so that it could easily fit with any other hospital information system infrastructure and organization. To do so a data gathering and formatting module (DGM) has to be designed to capture each new update of the patient record.

4.3 A Custom Patient Record for HAI monitoring

The initial research project to create a HAI detection system has made some assumptions with respect to the input format of the documents to be parsed among which we can notice: amonymization, time standardization, content coherence, etc. However the organization of data extracted from the hospital information system by the ad-hoc data gathering module is not so "clean". Therefore the gathering module has to generate a Custom Patient Record (CPR) compatible with what is expected by our HAI monitoring prototype.

This means that for one patient several collections of data are grouped together in one single custom patient record. This structure has the advantage to allow an analysis with specific content parsing rules and decisions rules for each type of treatment. Indeed elements presented in section "3.4 Causality" and "4 Results" show that there are differences between reports produced by different medical units.

The structure proposed for the custom patient records is the following:

- Patient ID
- Patient birth date
- List of files (coherent set of data for one specific treatment)
  - File ID
  - Date T0H provided by the hospital
  - Date T0D detected from texts
  - Reference Date T0
  - List of Documents
    - Document ID
    - Document type (e.g. medial unit)
    - Document Content (text)

4.4 Date of the origin

Another problem to be addressed is the proper detection of the reference date: T0. This is important as temporality management and reasoning for hypothesis validation is based on this reference date.

One solution could be to use the recording date that is associated with all information pushed in the hospital information system. However, after discussions it seems that this date cannot be trusted as a report is not immediately written and recorded after a given treatment. Therefore we have decided identify automatically the date of origin T0. For a given file (coherent set of data for one specific treatment for a given patient):

- A date T0H is provided by the data gathering module (DGM). This date is either the date of the main treatment (e.g. the surgery) if it is recorded in the hospital database, or the date when these documents have been recorded in the system.
- A date T0D is provided by an evolution of the anonymization tool designed for the research project. This tool parses the text content of the CPR to detect all dates or time reference. A date T0 is defined either through the detection of an explicit link between an event and a date (e.g. "… a knee surgery has been done on patient Mr X on June 6th 2011…") or through the comparison between the document redaction date and the closest date mentioned in the document.
- Then a separated decision module assign to the patient file the reference date T0. The decision can be taken according to the level of confidence assigned to each T0H and T0D date.

The decision algorithm should be tuned according to experiments performed on real patient records from the hospital information system.

4.5 Scalability and workflow

Another factor to be considered when delivering a monitoring system in a real hospital information workflow is the amount of information to be processed, the capacity of the system to handle the flow, and the amount of result data generated.

After discussions with people managing the hospital information system, we can anticipate a workload of 300 patient records updated per day with an average size of 30 KB per patient record. This makes approximately 9 to 10 MB of data to be processed per day. This can be easily processed by our system which is able to parse more than 2000 words per second, and even if it was not the case, the process is easily parallelizable. Therefore scalability is not an issue.

4.6 The decision module

The final aspect to be considered is the evolution of the monitoring system in a live environment. Decision heuristics have already been defined and evaluated on our research project. These rules are currently being improved to face the second and final evaluation. However this is done to cover the requirements of our initial project (orthopedic and surgery reports). In the context of a deployment in a real hospital the system should be customizable enough to allow its modification to address new types of bacteria, or new antibiotic drugs or even a modification of HAI classification criteria, but it raises some problems.

Terminologies can be easily updated if added expressions remain at the level of simple words. As soon as more complex expressions are concerned it implies a more important modification of the Part of Speech tagger that requires the expertise of a linguist. Furthermore adding new entity types will implies modification of the decision rules and a good understanding of their structure to avoid unpleasant side effects.

Finally modifying decision rules implies that results can change with respect to previous analysis. It is important to evaluate and control the impact.

5 Conclusion and Next Steps

We have presented in this paper the latest achievements on a research work to develop a Hospital Acquired Infection detection system from patient discharge summaries. Results of the 1st real evaluation of the system have demonstrated very interesting performances which has conducted us to consider an evolution of the system to plug it inside a real hospital information workflow.

This is a great opportunity to prove that an NLP based monitoring system can be used inside a hospital information workflow to improve patient safety.

Acknowledgments

The work presented in this paper comes from a 3 year project (ALADIN) started in 2009 and funded by the French *Agence Nationale de la Recherche* (National Research Agency - ANR) in the context TecSan program (*Technologies pour la Santé et l'Autonomie*).

References

Aït-Mokhtar S., Chanod J.P., (1997) *Incremental Finite-State Parsing.* In Proceedings of the Fifth Conference on Applied Natural Language Processing (ANLP'97), Washington March 31st to April 3rd, 1997, pp.72-79

Berrouane Y, Hagège C, Gicquel Q, Kergoulay I, Pereira S, Proux D, Darmoni S, Segond F, Metzger MH. *Preliminary evaluation of an automated detection tool for healthcare-associated infections, based on screening natural language medical reports.* Poster at the 21st European Congress of Clinical Microbiology and Infectious Diseases and the 27th International Congress of Chemotherapy. Milan from 7 - 10 May 2011.

Garner JS, Jarvis WR, Emori TG et al. *CDC definitions for nosocomial infections,*1988. Am J infect Control 1988;16 : 128-40

Hagège C; Marchal P; Gicquel, Q; Darmoni, S; Pereira S, Metzger MH. Linguistic and Temporal Processing for Discovering Hospital Acquired Infection from Patient Records. In the 2nd International Workshop on Knowledge Representation for Health Care (KR4HC-2010) - Workshop ECAI 2010, Lisbon, Portugal, August, 2010

Hagège C; Proux D; Gicquel, Q; Darmoni, S; Pereira S, Segond F. Metzger MH. *Développement d'un système de détection des infections associées aux soins à partir de l analyse de comptes-rendus d'hospitalisation.* In Proceedings of *TALN, Montpellier, France, June 27-July 1, 2011.*

H, S. E. Humphreys. *Prevalence surveys of healthcare-associated infections : what do they tell us, if anything?* Clin Microbiol Infect 2006; 12: 2-4

Marie-Hélène Metzger, Quentin Gicquel, Denys Proux, Suzanne Pereira, Ivan Kergourlay, Elisabeth Serrot, Frédérique Segond, Stéphan Darmoni. *Development of an Automated Detection Tool for Healthcare-Associated Infections Based on Screening Natural Language Medical Reports.* In American Medical Informatics Association conference, San Francisco CA, November 2009.

Towards Temporal Segmentation of Patient History in Discharge Letters

Galia Angelova
Institute of Information and Communication Technologies, Bulgarian Academy of Sciences (IICT-BAS)
Sofia, Bulgaria
galia@lml.bas.bg

Svetla Boytcheva
IICT-BAS and
University of Library Studies and Information Technologies
Sofia, Bulgaria
svetla.boytcheva@gmail.com

Abstract

This paper reports about ongoing work in automatic identification of temporal markers and segmentation of patient histories into episodes. We discuss the discourse structure of the *Anamneses* in Bulgarian hospital discharge letters and present experiments with a corpus of 1,375 anonymised discharge letters of patients with endocrine and metabolic diseases. Our IE prototype discovers 32,445 key terms in the corpus, among them more that 7,000 occurrences of drug names and about 7,500 occurrences of diagnoses. The temporal markers occur 8,248 times usually paired with tokens pointing the direction of time "forward" or "backwards". Temporal markers are identified with precision 84%, recall 57% and f-measure 67.9%.

1 Introduction

Medical informatics has made little progress in the temporal representation and reasoning tasks [1]. This is partly due to the complexity of the free text descriptions in clinical narratives where temporal information is presented. On the other hand, there is no agreement about the essence of clinical temporal models and the concepts and relationships that have to be taken into account. Research on temporal information interpretation is still in its embryonic stage according to [2]. The progress requires theoretic models as well as large training corpora of annotated medical texts which are expensive to construct.

This paper summarises work in progress on discharge letters structuring and experiments in automatic extraction of temporal markers. We deal with discharge letters in Electronic Health Records (EHR) in Bulgarian language. These letters contain predefined sections due to the general practice of structuring clinical notes into sections which dates back to the 60's and 70's of the last century as a result of centralised regulations[1]. Our present IE system, which analyses discharge letter texts, automatically identifies the *Anamnesis* (*Patient history*). This section contains a sketchy abstract, manually prepared by medical experts, who summarise the patient history in order to communicate it to another doctor. Explicit temporal markers designate the main phases in disease development, the main interventions and their effects. The unified text format is a motivation for its inclusion in methods for automatic episode recognition. Our main idea is to select a discourse structure theory and to try extracting pieces of information that have meanings as discourse units. The paper presents our first results in this direction.

Section 2 overviews related approaches. Section 3 considers the context of our work: the discourse structure of discharge letters and existing prototypes for section splitting and information extraction. Section 4 presents the evaluation of the current component for automatic extraction of temporal markers as well as our research agenda for building timelines of clinical events. Section 5 contains the conclusion.

2 Related Work

Research on temporal information processing is a relatively recent activity in biomedical NLP. Savova et al. [4] presents an annotation schema with temporal relations based on TimeML [5] and analyses the potential of TimeML tags as annotation tool for clinical narratives. The general objective is to build a temporal relation discovery component and a reasoner to create timelines of clinically relevant concepts. The authors consider five *Event* classes including *Occurence* (events that happened) and *State* (a condition or

[1] The list of sections in Bulgarian hospital discharge letters is published as a legal Agreement in the Official State Gazette, Article 190(3) [3].

Proceedings of the Workshop on Biomedical Natural Language Processing, pages 49–54,
Hissar, Bulgaria, 15 September 2011.

state, e.g. symptoms, descriptors and chronic conditions). The paper explicates important fine-grained characteristics of events and temporal relations in clinical texts which are related to linguistic units.

Five tags for marking up temporal information are suggested in [6]: *reference point*, *direction*, *number*, *time unit*, and *pattern*. The authors identified 254 temporal expressions in 50 discharge summaries and represented them using the suggested scheme. The inter-rater agreement was 75% which shows the complexity of temporal annotations even when simple tags are used.

Harkema et al. [7] presents an algorithm called ConText which identifies clinical conditions that are described in clinical reports: they can be *negated*, *hypothetical*, *historical*, or experienced by someone *other* than the patient. ConText infers the status of a condition with regard to these properties from simple lexical clues occurring in the context of the condition. The study deals with 4,654 annotations from 240 clinical reports: 2,377 annotated conditions in the development set and 2,277 annotated conditions in the test set. The evaluation summarises results obtained in a six-token window (*stw*) and end-of-sentence (*eos*) contexts. For *"historical"* condition, ConText achieves *stw* precision 78% and recall 70% as well as *eos* precision 77% and recall 79% across all report types that contain such conditions.

Hripcsak et al. model temporal information as contraint satisfaction problem [8]. Medical events from 231 discharge summaries are represented as intervals, and assertions about events are represented as constraints. Up to 151 medical events and 388 temporal assertions were identified per complete discharge summary. Non-definitional assertions were explicit (36%) or implicit (64%) and absolute (17%), qualitative (72%), or metric (11%). Implicit assertions were based on domain knowledge and assumptions, e.g., the section of the report determined the ordering of events. The source texts contained no instances of discontinuous temporal disjunction. The authors conclude that a simple temporal constraint satisfaction problem appears sufficient to represent most temporal assertions in discharge summaries and may be useful for encoding electronic medical records.

In our present work, we are mostly influenced by [6], which is attractive because of its relative simplicity, and [8].

3 Project context

The general objectives of our project are: *(i)* to develop a system for knowledge extraction from discharge letter texts and *(ii)* to design algorithms for searching conceptual patterns in the extracted clinical facts. Recognising the events and ordering them in timelines is an essential part of the project research agenda.

3.1 Materials

The discharge letters in our corpus contain sections which can be automatically recognised with accuracy 99.99% using their headers. Sometimes the structure is not strictly kept due to section merging, changing the section headers, skipping (empty) sections and replacing the default section sequence. Table 1 shows the percentage of discharge letters with available standard sections in the corpus of 1,375 EHRs.

Section title	*Discharge letters containing the section*
Diagnoses	100%
Anamnesis	100%
Past diseases	88.52%
Allergies, risk factors	43.56%
Family Medical History	52.22%
Patient Status	100%
Lab data, clinical tests	100%
Examiners' comments	59.95%
Debate	100%
Treatment	26.70%

Table 1. Formatting discharge letters according to centralised state regulations

Having the potential to identify automatically the *Anamnesis* section, which contains the patient history, we can plan further research tasks related to its temporal segmentation.

3.2 Methods

We have developed extractors of ICD-10 codes (the International Classification of Diseases, v. 10[2]) and ATC codes (the Anatomical Therapeutic Chemical Classification System[3]) from discharge letter texts [9, 10, 11]. The tool for diagnosis extraction assigns ICD-10 codes to Bulgarian nominal phrases designating disease names with 84.5% precision [9]. The drug extractor assigns ATC codes to Bulgarian drug names with f-measure 98.42% and achieves 93.85% f-

[2] http://www.nchi.government.bg/download.html

[3] http://www.who.int/classifications/atcddd/en/

measure for the dosage recognition [10]. Automatic extraction of "current treatment" is possible with highest accuracy for the drugs discussed in the *Anamnesis* (precision 88%, recall 92.45%, f-measure 90.17%, overgeneration 6%) [11].

Figure 1 shows the connections between the text (pre-)processing components that have been developed so far in our projects. The anonymised input discharge letter is split into section; after tokenisation and sentence splitting, the temporal markers are identified and episodes are tagged. Within the episodes, diseases and drugs are recognised by the existing tools. Other entities to be recognised in the episodes are *conditions* (symptoms, complains and status) and *treatment outcome*. The system works on a resource bank of medical nomenclatures, terminologies and linguistic resources. The output consists of annotated anamneses whit tagged episodes.

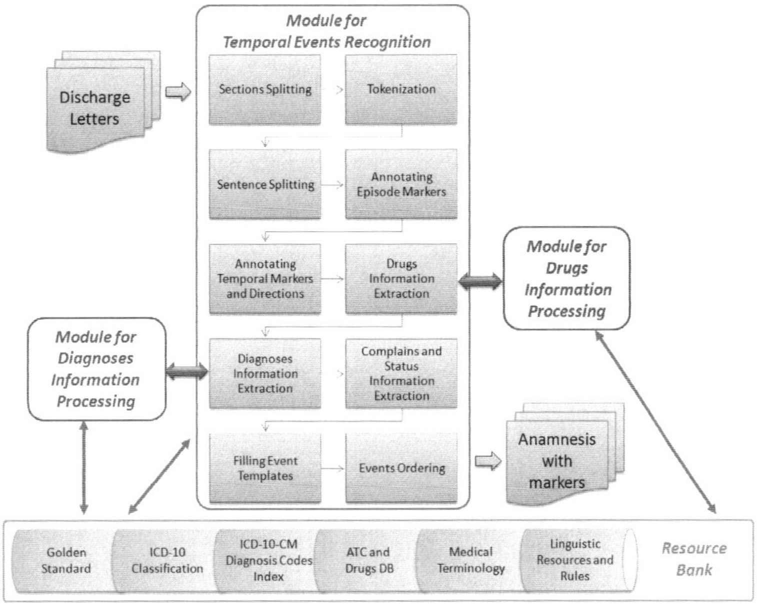

Figure 1. Text pre-processing components and extractors of diseases, drug names, and temporal markers

4 Extraction of temporal markers and ideas for structuring episodes

4.1 Evaluation results

We have performed experimental tests with 1,375 discharge letters where our IE prototype discovers 32,445 key terms or markers in the *Anamneses* (in average 23.59 per discharge letter). The distribution of these terminologies and temporal markers is presented in Table 2.

Temporal Marker	*Occurrences*	*EHRs*	*Avg*
drug names	7,108	1,213	5.86
diagnoses	7,565	1,292	5.86
complains	1,274	841	1.51
temporal	8,248	1,373	6.01
direction	8,249	1,374	6.01
Total	32,445	1,375	23.59

Table 2. Recognised entities in 1,375 discharge letters

Not surprisingly the temporal and direction markers occur as pairs, because direction markers point "forward" or "backward" from the time marker. The share of temporal and direction markers is significant (51% in total, see Fig. 2). These figures explicate the importance of temporal information for the case history description.

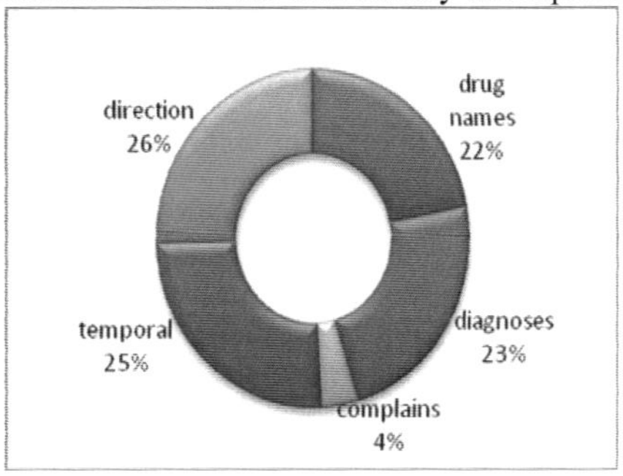

Figure 2. Percentage of temporal markers

The present recall in the recognition of the temporal markers in about 57% and the precision is 84% (f-measure 67.9%).

4.2 Segmentation into episodes

In medicine, an episode comprises all activities that are performed between the diagnosis of a disease and its cure (or stabilisation in case of chronic diseases). Studying various approaches to determine and annotate the granularity of temporal intervals, we view the patient history episodes as sets of events defined via the explicit temporal markers stated by the physicians who examine and treat the patients.

Most discharge letters in our corpus concern patients with endocrine and metabolic diseases diagnosed decades ago. In general only the major illness phases are discussed in the *Anamnesis* together with the treatment and medication changes. We consider below an example of a case history written in 2010:

Example 1. `Diabetes Mellitus diagnosed 5-6 years ago, manifested by most symptoms. At the beginning started treatment with Maninil only, afterwards in combination with Siofor. After few months the Maninil was replaced by Diaprel. Since October 2005 treated with Insulin Novomix 30 - 32E in the morning, 26E in the evening with diagnosed diabetic retinopathy. Complains of strong pains in the feet mostly at night.`

We believe that human experts declare explicitly the most important temporal markers which are sufficient (in their view) to adequately and unambiguously communicate the case history to another medical doctor. Therefore, we consider these markers as intentional signals for discourse segmentation. Our temporal model is framed using three tags suggested in [6]:

- *reference point*, *direction*, and *temporal expression*

plus additional tags needed for our project:

- *diagnoses or disorders,*
- *complains or symptoms,*
- *drugs/treatment applied* as well as
- *treatment outcome.*

There could be several diagnoses or symptoms enumerated in one episode as well as more than one drug correspondingly prescribed to the patient.

Let us consider the episodes in Example 1 which are defined by the explicit temporal expressions. Interpreting the text and ordering the temporal markers in a time progression scale according to the concrete moments, we construct the representation shown in Table 3. The conventional literal *'now'* denotes the speech/writing moment, in this case the moment of hospitalisation in 2010. Table 3 integrates the results of two extractors that were referred to in Section 3.2.

Ep1	Reference point	*Now* minus 5-6 years
	Direction	forward
	Temporal expression	5-6 years ago
	Diagnoses	Diabetes Mellitus E10
	Complains, symptoms	
	Drugs/Treatment	Maninil A10BB01
	Drugs/Treatment	Siofor 1 A10BA02
	Treatment outcome	
Ep2	Reference point	*(Now* – (5-6 years)) + few months
	Direction	forward
	Temporal expression	After few months
	Diagnoses	
	Complains, symptoms	
	Drugs/Treatment	Diaprel A10BB09
	Drugs/Treatment	Siofor 1 A10BA02
	Treatment outcome	
Ep3	Reference point	October 2005
	Direction	forward
	Temporal expression	Since October 2005
	Diagnoses	Diabetic retinopathy H36.0
	Complains, symptoms	
	Drugs/Treatment	Insulin Novomix 30 – 32E mon., 26E ev.
	Treatment outcome	
Ep4	Reference point	*Now*
	Direction	
	Temporal expression	
	Diagnoses	
	Complains, symptoms	strong pains in the feet
		.....

Table 3. Temporal segmentation with integration of automatic diagnoses and drugs extraction

Studying manually the discharge letters in our corpus, we think that the episodes resemble discourse segments as introduced in [12]. It seems reasonable to consider every temporal marker as a cue phrase signaling a new episode, because cue phrases express the intention of the writer to emphasize on major disease phases. In Example 1 and Table 3 we can also follow the *topical focus shifts* [13]: an episode might

- retain the topic of the previous one and contain references to the discourse entities in the previous episode. For instance, *episode 2* which was uttered immediately after *episode 1* refers to the entity Maninil introduced in *episode 1*;

- shift the topic and start discussion of another entity like e.g. *episode 3*.

Practical guidance of how to recognise segments is given in [14] where a discourse segment is viewed as a sequence of clauses that display local coherence. The following properties are listed as features that should hold within a segment:

(i) Resolution of references should be possible by techniques based on recency;

(ii) The time is fixed or there is a simple progression;

(iii) A fixed set of background assumptions is relevant to all clauses in the segment;

(iv) From intentional perspective, all the sentences in the segment contribute to a common discourse purpose, i.e. the same communicative goal motivates the writer to utter all clauses in the segment;

(v) From informational perspective, all the sentences in the segment are related to each other by some temporal, causal or rhetoric relations, i.e. all sentences and phrases combine together to describe a coherent event or situation.

Following these five properties, we might view the elementary *episode 1* and *episode 2* as a single segment because they discuss the progression of the diabetes, while *episode 3* is focused on the diabetes complication *Diabetic retinopathy* diagnosed in 2005. We also note that the clauses in *episode 1* and *episode 2* form a focus space where the diabetes progress in considered; in computational linguistics the focus spaces are organised in hierarchies and the preferred option is to open each topic only once.

These structural features are seen in most discharge letters that we have studied manually but we are far from final generalisation of our empirical observations. At the end we note that most temporal markers contain explicit references to time which help to construct the "elementary episodes". Only 0.014% of the temporal markers contain vague statements like "*long-term* diabetes" and only 0.01% use expressions like "since *several* years/weeks/days". The relative temporal markers are oriented in two directions:

- To the time marker of the immediately preceding previous episode, e.g. *since then*, *before that*, *after that*, etc. and
- To the moment *Now* when the discharge letter is written: like e.g. *few months ago*. Such episodes might elaborate a past event or period no matter that they are oriented according to the present moment.

The next task in our research agenda is to develop heuristics enabling to automatically position the temporal markers on a linear time scale with respect to the actual date of hospitalisation which is known in the Hospital Information System. In this way the episodes might be ordered in a sequence using a simple procedure which tries to calculate the actual date and constructs a list of linearly-ordered reference points. Moreover, applying the discourse coherence considerations, we could aim at semantic coupling of episodes which discuss the same topic (like *episode 1* and *episode 2* that are adjacent and display local coherence). This seems to be a challenge but one can aim to achieve it because adjacent episodes should be uttered in consecutive sentences dealing with the same entities. Figure 3 is obtained in this way. At present we only extract temporal markers and build "*elementary*" episodes of patient history phases.

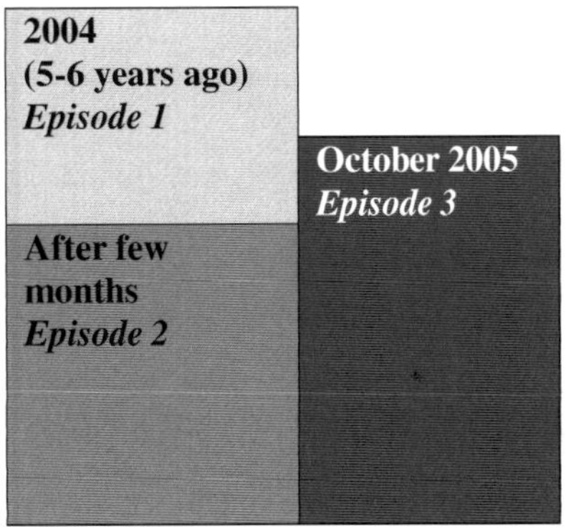

Figure 3. Grouping episodes into segments with local coherence for an *Anamnesis* written in 2010

Background medical knowledge might also help for the automatic grouping of adjacent episodes into intervals. For instance our system needs to perform semantic analysis of *episodes 1* and *2* in order to understand that Maninil A10BB01 was replaced by Diaprel A10BB09 in a few months. However, these two drugs belong to the same generic group (which is seen by the ATC code) and cannot be taken together, and this medical default might support the semantic analysis of the textual description. This fact is another hint that *episode 2* elaborates *episode 1* and presents further description of the treatment related to diabetes. Therefore, it is easy to find informational signals that *episode 2* belongs to the local context of *episode 1*.

5 Conclusion

This paper presents work in progress aiming at the automatic segmentation of episodes in the patient history (usually events or periods) as described in discharge letters.

So far, we have found no case of temporal ambiguity which prohibits the manual annotation (although it is based on human interpretation of the text semantics). Further research needs to be done for designing a heuristic strategy of how to "glue" adjacent groups of clauses together because they display local coherence properties. To solve such a task our system needs to understand which episode elaborates the previous one. The present paper reports first findings in this respect.

Acknowledgments

The research work presented here is supported by grant DO 02-292 "Effective search of conceptual information with applications in medical informatics", funded by the Bulgarian National Science Fund in 2009-2012.

The authors are grateful to the paper reviewers for the useful comments and numerous suggestions.

References

1. Zhou L., C. Friedman, S. Parsons, and G. Hripcsak. *System architecture for temporal information extraction, representation and reasoning in clinical narrative reports.* In Proc. AMIA Ann. Symp. 2005, pp. 869–873.
2. Zhou L. and G. Hripcsak. *Temporal reasoning with medical data - a review with emphasis on medical natural language processing.* J. Biom. Informatics 2007, 40(2), pp. 183-202.
3. *National Framework Contract* between the National Health Insurance Fund, the Bulgarian Medical Association and the Bulgarian Dental Association, Official State Gazette №106 (2005), updates №68(2006) and №101(2006), Sofia, Bulgaria, http://dv.parliament.bg/.
4. Savova, G., S. Bethard, W. Styler, J. Martin, M. Palmer, J. Masanz, and W. Ward. *Towards Temporal Relation Discovery from the Clinical Narrative.* In Proc. AMIA Annual Symposium 2009, pp. 568–572.
5. Sauri R., J. Littman, B. Knippen, R. Gaizauskas, A. Setzer and J. Pustejovky. *TimeML annotation guidelines*, Version 1.2.1, 31 January 2006. Available online at http://www.timeml.org/site/publications/timeMLdocs/annguide_1.2.1.pdf.
6. Hyun S., S. Bakken and S.B. Johnson. *Markup of temporal information in electronic health records.* In Stud. Health Technologies and Informatics Vol. 122, 2006, pp. 907-908.
7. Harkema, H., J. Dowling, T. Thornblade, and W. Chapman. *Context: An Algorithm for Determining Negation, Experiencer, and Temporal Status from Clinical Reports.* J Biomed Inform. 2009 42(5): 839–851.
8. Hripcsak G., L. Zhou, S. Parsons, A. K. Das, and S.B. Johnson. *Modeling electronic discharge summaries as a simple temporal constraint satisfaction problem.* JAMIA 2005, 12(1), pp. 55-63.
9. Tcharaktchiev, D., G. Angelova, S. Boytcheva, Z. Angelov, and S. Zacharieva. *Completion of Structured Patient Descriptions by Semantic Mining.* In: Koutkias et al. (Eds.), *Patient Safety Informatics - Adverse Drug Events, Human Factors and IT Tools for Patient Medication Safety*, Stud. Health Techn. Inform. 166, 2011, pp. 260-269.
10. Boytcheva, S. Shallow Medication Extraction from Hospital Patient Records. In: *Koutkias* Koutkias et al. (Eds.), *Patient Safety Informatics – Adverse Drug Events, Human Factors and IT Tools for Patient Medication Safety*, Stud. Health Techn. Inform. 166, 2011, pp. 119-128.
11. Boytcheva, S., D. Tcharaktchiev and G. Angelova. *Contextualization in automatic extraction of drugs from Hospital Patient Records.* In A. Moen et al. (Eds.) *User Centred Networked Health Case*, Proc. of MIE-2011, the 23th Int. Conf. of the European Federation of Medical Informatics, Stud. Health Techn. Inform. 169, 2011, pp. 527-531.
12. Grosz, B. and C. Sidner. *Attention, Intention and the Structure of Discourse.* Computational Linguistics 1986, 12(3), Reprinted in RNLP.
13. Grosz, B. *The representation and use of focus in a system for understanding dialogues.* IJCAI 1977, pp. 67-76, Reprinted in RNLP.
14. Allen, J. *Natural Language Understanding*, 2nd Edition, Benjamin/Cummings, 1995.

Author Index

Angelova, Galia, 49

Boytcheva, Svetla, 11, 49
Bretonnel Cohen, Kevin, 1

Darmoni, Stefan, 43

Georgiev, Georgi, 27
Gicquel, Quentin, 43

Hagège, Caroline, 43
Hassel, Martin, 3
Henriksson, Aron, 3

Kokkinakis, Dimitrios, 35

Metzger, Marie-Hélène, 43

Nakov, Preslav, 27

Pereira, Suzanne, 43
Popov, Borislav, 27
Proux, Denys, 43

Segond, Frédérique, 43
Sokolova, Marina, 19

Zhivkov, Valentin, 27